VCR Connections Guide

VCR Connections Guide

TV Accessory Hookup Made Easy

The Howard W. Sams Engineering Staff
Louis E. Frenzel, Jr., Consulting Author

HOWARD W. SAMS & COMPANY
A Division of Macmillan, Inc.
4300 West 62nd Street
Indianapolis, Indiana 46268 USA

FIRST EDITION
FIRST PRINTING—1988

International Standard Book Number: 0-672-22648-0
Library of Congress Catalog Card Number: 88-62229

Acquisitions Editor: *James Rounds*
Manuscript Editor: *Susan Pink, TechRight*
Illustrator: *T. R. Emrick*
Cover: *Bruce Dean—Dean Johnson Design*
Indexer: *Don Herrington*
Composition: *Impressions, Inc.*

Printed in the United States of America

Trademark Acknowledgments

Contents

Hookup Finder

Preface

VCR Connections Guide: TV Accessory Hookup Made Easy explains in a simple way how to interconnect VCRs and other video accessories to a television. Although the user's manuals supplied with VCRs and other accessory units generally provide adequate hookup information for simple connections, they do not cover special cases involving two or more video devices. Today, it is common for consumers to have not only a VCR, but also a cable TV selector box, a satellite TV receiver, a stereo decoder, or some other accessory.

When multiple devices are used with the TV set, complications occur. Specifically, the cabling and interconnections become complex and tricky. Making multiple interconnections usually increases the functionality of your system, but it can also introduce limitations and even degrade the picture and sound if you are not careful. Trying to make such interconnections without specific instructions can be frustrating and disappointing. It is maddening to invest in expensive video devices and not be able to make them work together harmoniously.

Of course, you can always call a service technician to install and connect your equipment. But with the price of service calls, you no doubt wish to avoid this option. Another alternative is to get a knowledgeable friend to help. This may be tougher than you think because even technical experts get confused.

Your best bet is this book:

- It costs much less than a service call.
- It contains all the data you need to hook up video equipment.

- It helps you set up a working system for your current equipment.
- It also helps you make necessary changes when you add to the system in the future.

This book is possible through the combined efforts of the author and the Howard W. Sams and Company Engineering staff. With new video equipment described and many hookup arrangements included, the book covers the most common combinations of available accessories. Better still, this information is organized to help you find the hookup you need . . . fast.

VCR Connections Guide also includes information about how accessories work and why they must be connected in certain ways. By learning the simple theory behind interconnection, you will be able to figure out how to create a unique system in the future. Even though the system may not be described in this book, the information here will help you figure out exactly what you need to do and how to do it.

The world of video is truly exciting today. There are so many nifty and affordable video goodies to buy for your TV set to enhance your viewing and listening enjoyment. But to make them all work together requires some skill. This book makes it possible for you to hook up virtually anything without being an electronic whiz.

Good luck and best wishes.

Lou Frenzel

1 Getting Started

It used to be so simple. You bought a TV, then watched it. The most complicated part was to connect an outside antenna, if necessary.

Then along came more VHF channels. UHF channels were introduced later, giving you even more viewing choices. Still it was pretty simple. You probably had to add a UHF antenna or switch to a combination VHF/UHF antenna to get decent reception.

Remote control was next. That helped make channel selection fast and easy.

The next big developments were cable TV and VCRs. Suddenly you had more channels plus premium movies, sports events, and other programs. The VCR let you play prerecorded movies and other videos and also gave you the ability to record programs off the air. The VCR let you shift your viewing to a more convenient time and save shows of interest. Videodisc players came along for a brief period, giving you one more source of program materials. Mercifully, these didn't last.

Today, you are bombarded by more electronic products to attach to your TV set. There are video games, personal computers, and stereo adapters. Stereo TV is common now, and there is a strong incentive to tie your TV set into your hi-fi system for even better sound.

Satellite TV receivers let you pick up TV signals directly from satellites used by the networks and cable TV companies. Video cameras are another accessory and are becoming more affordable.

What you have today is not just a TV set, but a full video system made up of many different pieces. Your TV set is the basic viewing device to which you can attach many additional signal sources and accessories. All those boxes connected to your TV set let you customize the system. The big problem is hooking up all these devices.

If all you're going to do is attach a VCR or cable TV box to your set, you can easily get by with your equipment instruction manuals. But if you want a VCR and a cable TV box and any more accessories, get ready for some electronic fun. Although all these devices are compatible, trying to interconnect them so they all work properly with your TV is a challenge because it requires a maze of wires and equipment manuals don't tell you how to hook up more than one device.

Well, fret no more. This book will help you resolve any interconnection problems you may have. Better still, it will serve as a reference guide if you modify your system later by buying new accessories or replacing dated units.

How to Use This Book

This book is written for two types of people. It first addresses you no-nonsense types who want to get all your equipment hooked up right now. Your main objective is to start using your system to the fullest as soon as possible. The last thing you want to do is read a whole book on the theory of television or the concepts of interconnection. You want to know *how* but not *why.*

If you are this type, you can go directly to chapter 4 and figure out exactly what you want to do. Also scan the table of contents for chapters 5, 6, and 7 to see which hookup applies. Then refer to the appropriate hookup in chapter 5, 6, or 7, and you can immediately implement it. Instant gratification. Then put this book on your shelf to use later when you plan changes in your system.

One word of advice. If you run into trouble, refer to chapters 2, 3, or 8. Most likely, the answer will be in one of those chapters.

This book is also for you enthusiast types who would like to know what you are doing. Video may be a hobby and you've got more than a passing interest in electronics. This book is for you too because it tells you *how* and *why.* It ensures that you get everything hooked up and working as you want it. But it also explains each component, accessory, and interconnection. You learn how the equipment operates and why it is connected in a special way. By understanding how the equipment works, you can figure out new hookups yourself. If this is the route you wish to take, just read the chapters in sequence, ending with the appropriate hookup in chapter 5, 6, or 7. Chapter 2 discusses the theory of operation of the various units. Chapter 3 introduces all the related interconnection hardware. Chapter 4 helps you prepare for and select a hookup. It covers basic assumptions and definitions and gives you initial guidelines in narrowing your choice of hookups.

Finally, chapter 8 covers maintenance and troubleshooting. It explains things you can do periodically to keep your system running. It also tells you how to identify problems, isolate faults, and make simple repairs.

One final suggestion. While you are reading, you may encounter a word or term you don't understand. Although we define some terms as we present the material, we assume you do know the meanings of basic terms. If you run into trouble, use the glossary, which contains concise definitions of all key words.

What You Will Need

Before you begin, find all your equipment manuals and keep them handy. They will serve as backup references. You may need them to look up special features, operational details, or unique capabilities and limitations.

Also check to see if you have the necessary tools: a small flat blade screwdriver and a knife to strip wires.

Now, let's get started.

Where to Get More Information

Books:

1. *Connecting* by Tom Meyers, Harper & Row Publishers, 1987.
   Detailed connection of video and audio systems.
2. *Electronic Connections: Home and Car Entertainment Systems* by Martin Clifford, Prentice-Hall Inc., 1987.
   In-depth coverage of all audio and video system hookups.

Video Tape:

3. *The Cable TV-VCR Hook-Up Guide,* Naczinski & Associates Inc., Los Angeles, CA, 1987.
   Step-by-step, do-it-yourself hookup instructions in video format emphasizing cable converter box connections with TV sets and VCRs.

2 Video Equipment

A home video system consists of a TV set, one or more signal sources, appropriate accessories, and related cabling. Figure 2-1 shows the basic configuration. The TV set is the heart of the system because it provides the final picture and sound output. Everything else is a signal source or an accessory that provides enhancement or benefit.

A signal source, as its name implies, is some place that a video signal comes from. It can come from an antenna or a cable TV box, VCR, video game, computer, videodisc, or camera. These multiple sources of signals are what makes your video system so flexible. With multiple sources, you can choose among many viewing options and choose when as well as what you watch.

You can see from figure 2-1 the problem in multiunit video systems. Each signal source must be connected to the TV set. Basically, all of the signal sources are vying for the attention of the TV set. Therefore, you must have some way of selecting which signal source you want. The box labeled *interconnection system* in figure 2-1 represents all the cables, connectors, and accessories that permit all the input sources to share the TV set. The contents of that box are the main subject of this book.

This chapter summarizes the main features and basic operation of all major elements in the video system illustrated in figure 2-1.

Figure 2-1 A modern video system.

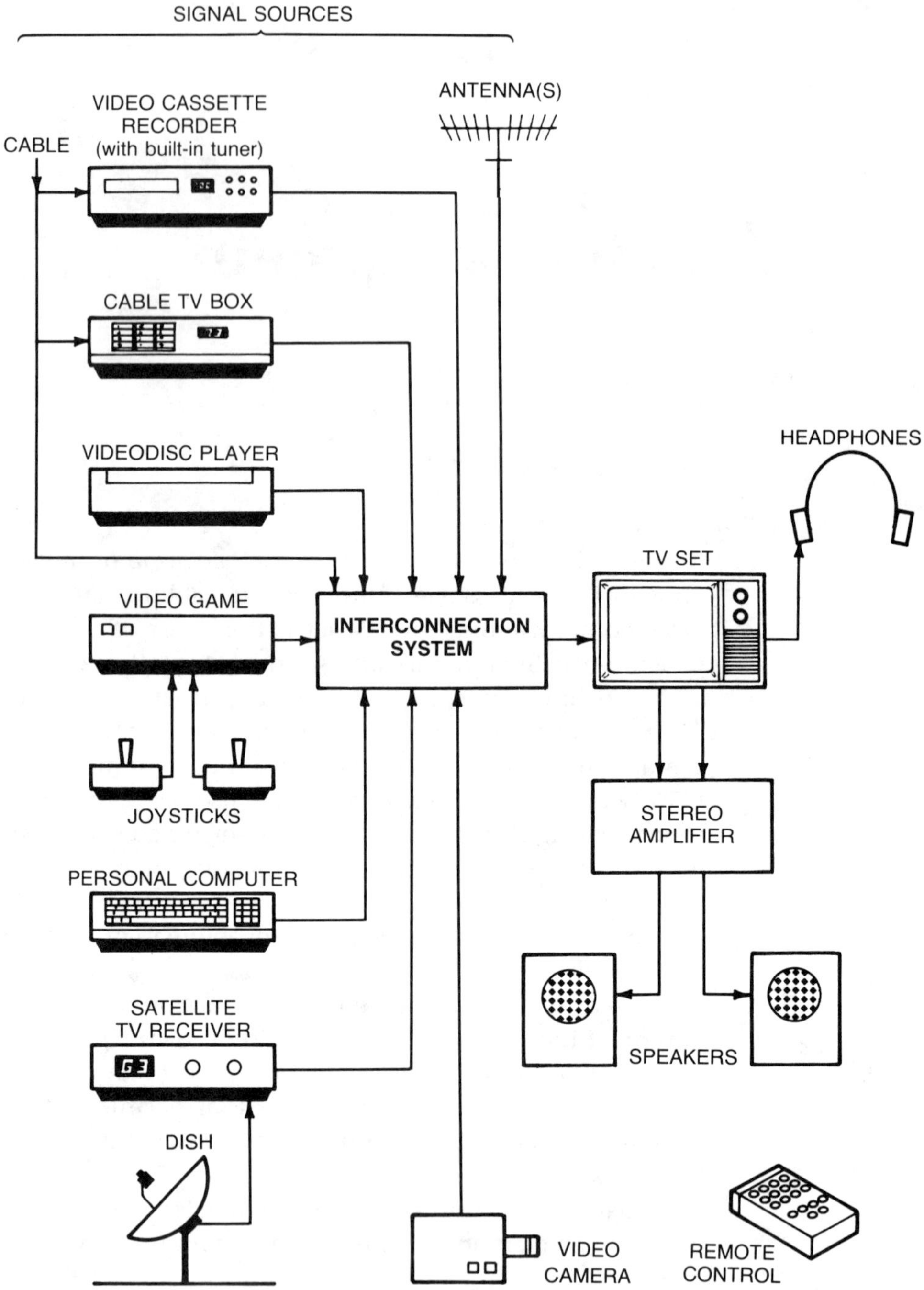

Video Signals

It will help to know what kind of signals you will be working with in the system. There are three basic types: video, audio, and RF. Video and audio signals are the basic picture and sound signals, respectively. The RF (radio frequency) signal is a high-frequency carrier that transports the video and audio to their ultimate destination.

All original video signals are derived from a video camera. The camera contains a vidicon tube or semiconductor sensor called a charge-coupled device (CCD) that converts the light in a scene into an electronic signal. This is called the video. Video signals range in frequency from direct current (DC) to over 4 MHz.

The audio signal is derived from a microphone. Two or more microphones are used in stereo TV to produce left and right signals. The audio signals are typically in the 50 Hz to 15 kHz frequency range.

The video and audio signals are not transmitted directly over the air or through a cable. Instead, they are impressed upon a higher frequency signal called a carrier. The carrier is an RF signal transmitted through free space or a cable.

The process of impressing the video and audio on a carrier is called modulation. During modulation, some aspect of the carrier wave is modified by the modulating signals. For example, the video signal varies the amplitude, or signal level, of the RF carrier. This is called amplitude modulation (AM). The sound varies the frequency of an RF carrier. This is known as frequency modulation (FM).

The main frequency of the carrier is selected to put the signal into a particular portion of the frequency spectrum. Each different carrier frequency is given a channel number. For example, TV signals in the 174 to 180 MHz range are designated as channel 7. A tuner in the TV set, cable box, or VCR selects the desired channel.

To recover the original video and audio signals, the RF carrier must be demodulated. Demodulation or detection is accomplished by a circuit that strips away the carriers. This leaves only the original video and audio signals, which are then displayed or output.

In any video system, you will deal with all three types of signals. Each is handled differently. You need to know which inputs or outputs on each device are RF, video, or audio to hook them up properly. The video and audio inputs or outputs are usually labeled as such. The RF inputs are sometimes labeled Antenna or Cable. On some sets, the inputs are labeled VHF and UHF.

TV Set

Let's begin our discussion of the various system components with the unit that anchors the entire system: your TV set. It is the starting point for everything else. Basically, your TV set is the system output. It is a presentation device. It displays the picture on the screen of the cathode ray tube (CRT) and outputs the sound to one or more speakers.

The TV set is also a signal selector and signal source. That is, it contains a tuner to select which of many RF inputs applied to its antenna terminals will be displayed. The antenna terminals are the TV set's inputs. RF TV signals from an antenna or any one of the many sources listed previously can be applied to the antenna terminals.

Figure 2-2 shows the basic block diagram of a TV set. The tuner selects the channel to be displayed. The resulting signal is then applied to the video and audio sections. The video section demodulates the RF signal and displays the picture on the CRT. The audio section demodulates the RF and presents the recovered sound to the speakers.

The only inputs on most TV sets are the VHF (very high frequency) and UHF (ultrahigh frequency) antenna terminals. These accept RF signals. Therefore, the desired video and audio signals must previously have modulated a carrier on a specific channel. TV broadcasts are of this form. Video signals from other sources such as a tape or camera must first be used to modulate a carrier before they can be viewed on the TV set. VCRs have a built-in modulator that puts the video signal from the cassette tape onto an RF carrier on either channel 3 or 4, which is then connected to the TV set's VHF antenna terminals.

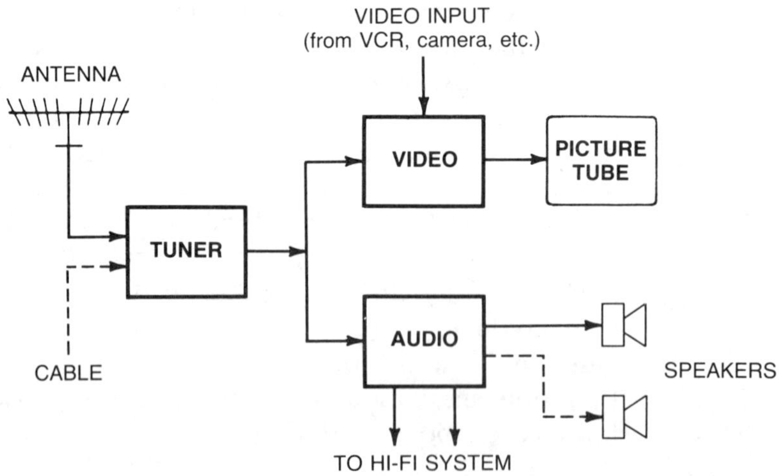

Figure 2-2
Block diagram of a TV set.

Many newer TV sets feature a video input jack, however. This allows video signals to be connected directly to the CRT display circuits. When the TV set is used this way, it is usually called a video monitor or monitor. The tuner circuits are bypassed completely. This usually improves picture quality because it eliminates all modulation and demodulation circuits, which can degrade performance. A switch on the back of the set selects either the internal tuner or the external video input as the source to be displayed.

TV sets with separate video inputs also usually have an audio input jack. In this way, the audio accompanying the video can be routed through the TV audio circuits and speakers. Check your TV instruction manual for using the video and audio inputs.

Some new TV sets have a tuner that is cable-ready. This means the tuner is capable of receiving signals directly from a cable TV input cable without the use of a cable converter box. Many of these sets receive well over one hundred channels. If you have such a set, you may be able to use it in lieu of the converter supplied by the cable company.

But many cable companies use special or unusual channel assignments, thereby making channel selection on a cable-ready set a problem. Further, scrambled or encrypted channels such as HBO and other premium cable channels cannot be received on a cable-ready set. The separate converter box is necessary to unscramble and decode these protected signals.

Figure 2-3 shows the symbol we use to represent the TV set in our hookup diagrams. The typical rear panel connections are clearly illustrated. There are two types of connections, screw terminals and coax connectors. The screw terminals accept twin lead cable, and coax connectors accept coaxial cable. You will learn more about these in chapter 3. The VHF input can come from a 75-ohm coax cable source or a 300-ohm twin lead source. The source can be a TV antenna, a VCR, or some other device. Not all TV sets have the VHF coax connector; they have screw terminals instead. Some sets have both screw terminals and coax VHF inputs. The UHF input is 300-ohm screw ter-

Figure 2-3 Symbol for television used in the hookups. The VHF and UHF terminals are for the antenna or RF outputs.

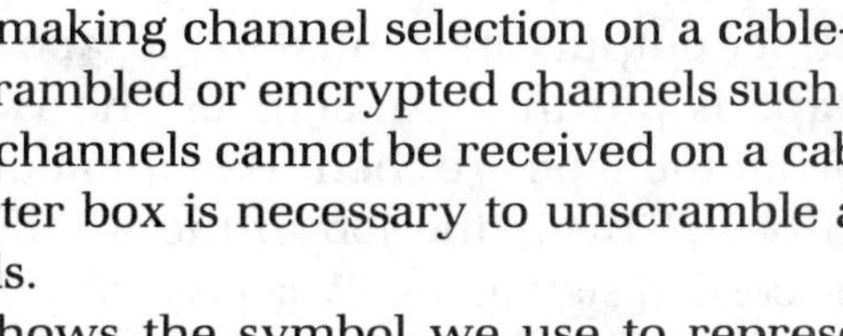

minals only. The video input is 75-ohm coax only. Only the newer sets designated as monitors have a video input. More about cable types and impedances later.

Video Cassette Recorder

The video cassette recorder (VCR) records or plays back video signals on ½-inch tape cassettes. A VCR is like a TV set without a picture tube. It has a built-in tuner, so it picks up all VHF and UHF TV channels by itself. Many newer VCRs contain cable-ready tuners. The VCR recovers the video signal and records it or passes it along to the TV set for display. The VCR must be used with a TV set or monitor if you want to see the picture.

Figure 2-4 shows a simplified block diagram of a VCR. The tuner selects a TV channel and demodulates the TV signals producing the original video and audio.

The tape unit shown in figure 2-4 records or plays. In record mode, tuner outputs are routed to the tape unit. In play mode, a prerecorded tape is put in the machine. The video and audio signals recovered from the tape are then used to modulate an RF carrier set to channel 3 or 4. This is the job of the RF modulator in figure 2-4. To view a recorded signal, the TV set is used. The TV set picks up only RF signals. But the recorder output is video and audio. So the RF modulator puts the video and audio on an RF carrier set to channel 3 or 4. This is applied to the TV set for normal reception.

All VCRs also have separate video and audio inputs and outputs. These bypass the tuner and provide direct connections to the tape unit inputs and outputs. Signals applied to the video/audio inputs are recorded by the tape unit. The video/audio outputs of the tape unit can be connected directly to the video and audio inputs on the TV

Figure 2-4 Block diagram of a VCR.

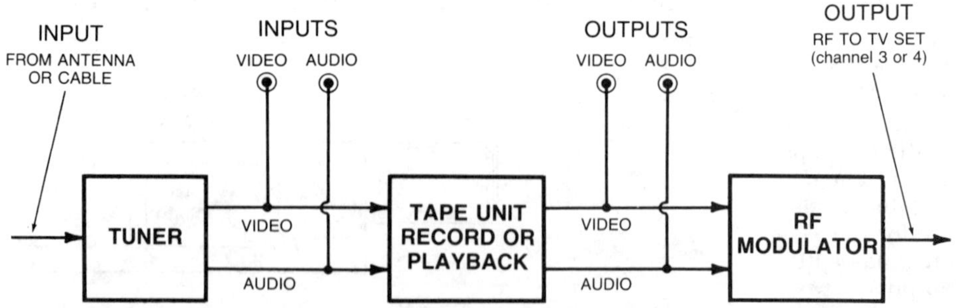

set or monitor, bypassing the RF modulator and the TV set tuner. This usually provides higher picture and sound quality.

A key part of the VCR is the TV/VCR switch. This is a front panel pushbutton that controls internal electronic switches that route the various input and output signals as required to perform one of several functions. This control may be labeled Video/TV or TV/Video. Check your VCR manual.

When the TV/VCR switch is in the TV position, the antenna or cable input signals applied to the VCR tuner are also routed to the VCR output terminals, then to the TV set antenna input terminals. The VCR is bypassed, and the TV set uses whatever input signals are available. The TV/VCR switch is automatically put in the TV position when the VCR is off and when it is first turned on.

If the TV/VCR switch is set to VCR, the input signals from the antenna or cable drive the VCR tuner. The VCR tuner then selects a channel to record or display depending upon which VCR operation you select.

The internal electronic switches operated by the TV/VCR button are also controlled by the VCR circuits themselves depending upon which VCR mode of operation is selected. The basic modes are stop, record, and play. Let's see what happens in each mode.

In stop mode, the VCR is on but nothing is happening. Figure 2-5 shows the effect of the TV/VCR switch in each position. Look at figure 2-5A. The switches shown may not actually be set up like this in the VCR, but the configuration is correct to illustrate the concept. In the TV position, the antenna or cable input to the VCR is connected to the VCR output and the TV antenna input. With this arrangement, you use the TV set as usual, selecting the desired channel with the TV set controls.

If you set the switch to VCR, the arrangement in figure 2-5B applies. The VCR antenna or cable input is connected to the VCR tuner. The tuner outputs are connected to the RF modulator, which in turn sends a channel 3 or channel 4 signal to the TV set. With this configuration, the TV set displays the channel the VCR is tuned to. The TV set must be set to channel 3 or 4. You select the channel with the VCR. Incidentally, if you are rewinding or fast forwarding the tape, the effect of the TV/VCR switch is the same as that in stop mode.

In record mode, the TV/VCR switch determines whether you watch what you are recording or something else. Refer to figure 2-6A. With the switch in the TV position, the VCR antenna or cable input is connected to the VCR tuner and directly to the VCR output terminals, bypassing the RF modulator. You use the VCR channel selection con-

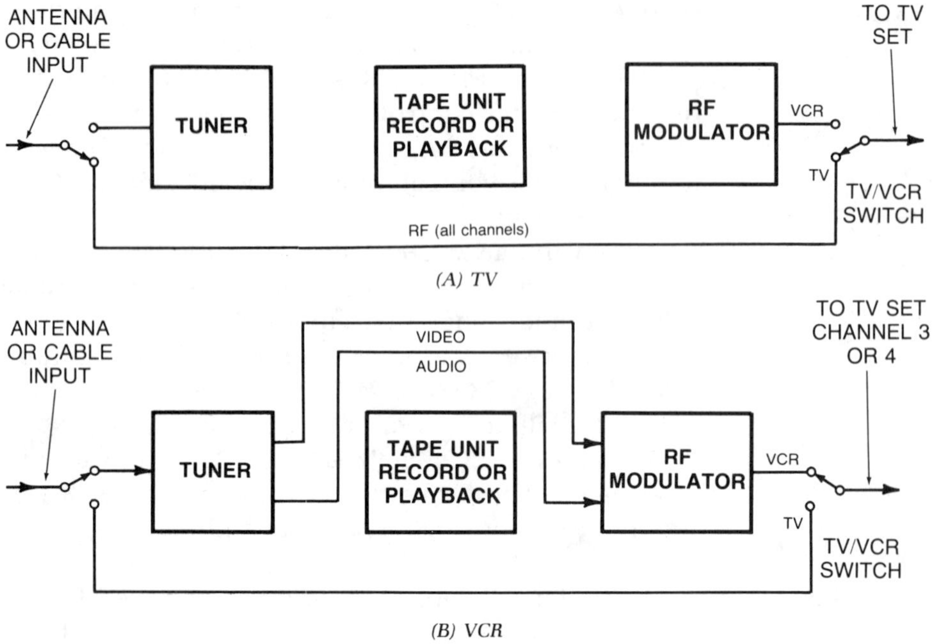

Figure 2-5 Stop mode operation in the VCR.

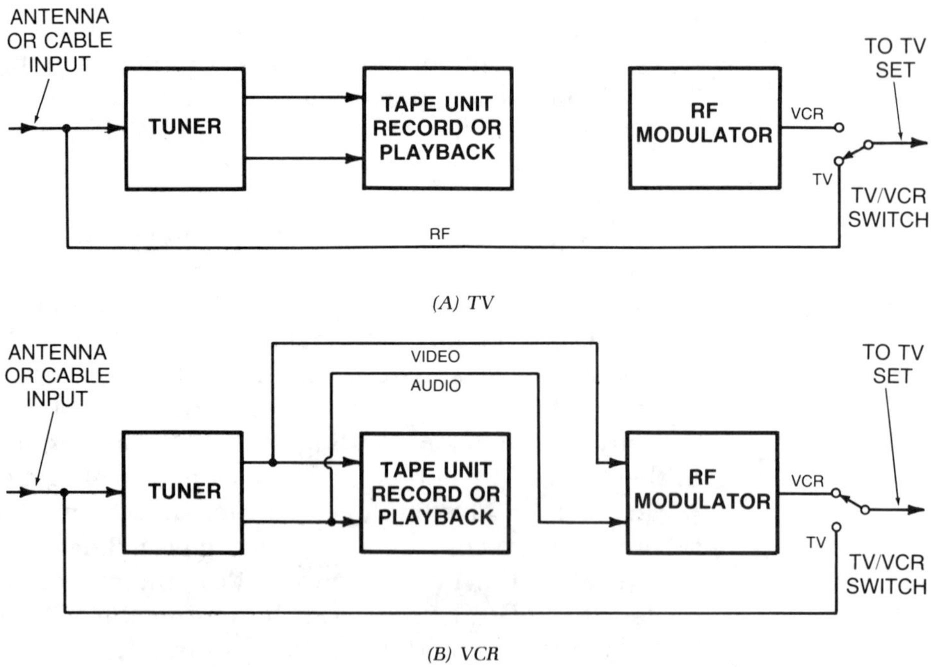

Figure 2-6 Record mode operation in the VCR.

trols to choose the channel to be recorded. But you use the TV set tuner to select a channel to view. This lets you watch one channel while you record another.

Now refer to figure 2-6B. If the TV/VCR switch is set to VCR in record mode, the VCR inputs also get connected to the VCR tuner, which selects the channel to be recorded. The tuner output is sent to the recorder and the RF modulator. With the TV set tuned to channel 3 or 4, you will watch the channel being recorded.

In play mode, the TV/VCR switch has no effect. The internal VCR circuits take over and switch the tape unit outputs to the RF modulator. The VCR antenna or cable inputs are disconnected. In play mode, only one thing happens. The tape being played is displayed on the TV. As before, the TV must be set to channel 3 or 4. See figure 2-7. In some VCRs, a high-speed search mode is available. In this mode, you will also see the tape output on the TV but at a higher speed.

Why Channel 3 or 4?

Why do you have to set your TV set to channel 3 or 4 to view the output of the VCR? The reason is the RF modulator in the VCR (or any RF modulator) can be tuned to either of these channels. The RF modulator contains a carrier oscillator whose frequency can be selected with a switch. In one position, the switch tunes the carrier to channel 3; in the other position, it tunes the carrier to channel 4. This switch is usually on the back of the VCR.

Channels 3 and 4 were selected because all TV sets cover them, and oscillators are cheap and simple to make for these frequencies.

To determine whether you should use channel 3 or channel 4, simply note if you have a nearby local station on either channel. If so, select the other channel. For example, if you have a channel 4 station in your area, select channel 3. By doing this, you eliminate the possibility that the strong local TV signal will interfere with the RF modulator output or vice versa. If you have neither channel 3 nor channel 4 stations in your area, select either channel.

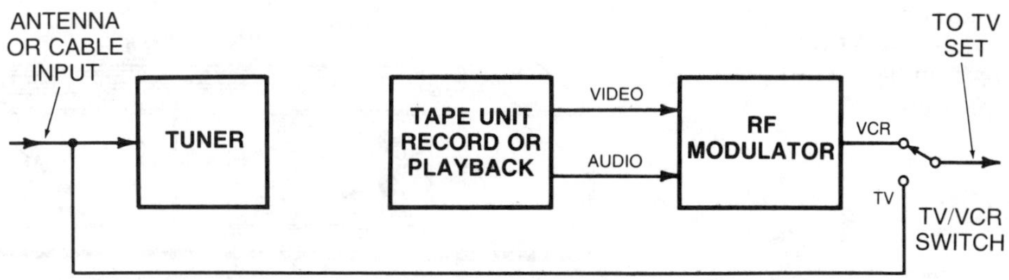

Figure 2-7 Play mode operation in the VCR.

Figure 2-8 shows the typical connections on the rear of a VCR. There are inputs for both VHF and UHF antennas. The VHF input is a coax connector, and the UHF input is two terminals for 300-ohm twin lead. If you have cable TV, the output of the cable TV converter box connects to the VHF input connector.

The VCR also has two outputs, again both VHF and UHF. These go to the VHF and UHF antenna terminals on your TV set. In some VCR modes, the VCR inputs are routed directly to these output terminals. In other modes or depending upon the position of the TV/VCR switch, the output of the RF modulator in the VCR is connected to the VHF output connector on the back of the VCR. The RF modulator input (VCR tuner or tape player) is sent to the TV set and displayed if the TV is tuned to channel 3 or 4.

VCRs also have direct video and audio inputs and outputs. These are available on RCA-type phono plugs, as figure 2-8 shows. The inputs can come from a camera or another VCR. The outputs can go to a TV set, monitor, or hi-fi stereo amplifier. Many newer VCRs feature stereo sound decoders. If your TV set doesn't have stereo capability, you can enjoy stereo by using your VCR for channel selection and decoding. Connect the stereo audio outputs of the VCR to your hi-fi stereo system for best performance.

The VCR is really a key component of any video system. Besides letting you record signals or play back prerecorded tapes, it also includes a separate tuner, RF modulator, and a switching system. These extra components give you added flexibility in selecting signal sources.

Cable Converter

Although a TV set and a VCR form the core of most video systems, many additional accessories have become popular. One of the most widespread new accessories is a cable converter box. Approximately half of the households in the United States are already served by cable

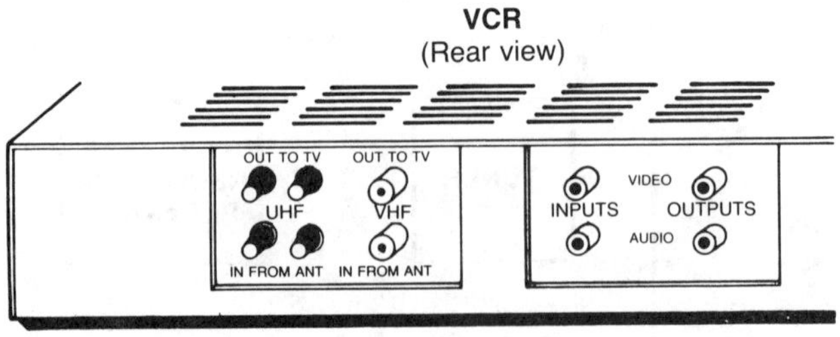

Figure 2-8 Symbol for VCR used in the hookups.

TV systems. In addition to providing high-quality signals of standard over-the-air channels, cable companies provide a wide range of additional programming, including premium movie channels, news, sports, and entertainment. The cable converter box supplied by the cable company brings these signals to your TV set.

The cable or cables that run from the cable company to your house supply a range of TV signals. To convert these to a signal compatible with your TV set, the cable converter box is required. If you have a cable-ready TV or VCR, you might not need a cable converter box. But if you plan to use any scrambled premium channels, you need the converter box to descramble them.

The cable converter is essentially a TV tuner with several important features. First, it often provides many more channels than the average TV set can receive. In some systems, well over one hundred channels are available. Second, the cable converter box often has descrambling circuitry for decoding the specially protected premium channels. A typical cable box is shown in figure 2-9. In most cases, channel selection is done by pushbutton switches. In some smaller systems, one pushbutton is available for each channel. In larger, more sophisticated boxes, a standard keyboard is provided to enter the channel number. Remote controls are typical in new units.

The incoming cable, which is typically 75-ohm coax, is connected to the back of the cable box. Most systems use one cable. But larger systems use two coaxial cables, one for standard channels and the other for protected premium channels.

The output of the cable converter box is also a coaxial connector that contains the selected cable channel, which modulates an RF carrier on channel 3 or 4. The cable converter contains a built-in RF

Figure 2-9 A cable converter box.

modulator that generates the output signal that is compatible with the TV set. The cable converter output is connected to the VHF antenna input terminals on the TV set. Alternatively, it can be connected to the VHF antenna input on the VCR if one is used in the system.

There are many different kinds of cable converter boxes. Following is a summary of the most common types.

Block Converter

A block converter converts a selected band of stations to another band of stations. It is usually used for tuning on a standard TV receiver. In the hookups for chapters 5 and 6, block converters always appear as illustrated in figure 2-10.

Up/Down Converter

An up converter changes the range of received frequencies to a higher range. A UHF up converter is a converter box that converts the entire cable band to a selected range of UHF channels. Cable channels are often converted to a UHF band so that a TV remote control can be used to select cable channels.

In most hookups, the converter box changes the selected channel to channel 3 or 4 TV output. With the up converter, the whole band of cable channels is changed to UHF channels. As a result, a TV remote control can select cable stations by tuning UHF channels.

The down converter uses the same principle to change a range of received frequencies into a lower viewable range. In the hookups for chapters 5 and 6, the down converter always appears as illustrated in figure 2-10.

Converter-Selector

A converter-selector is a converter box with a channel selector on the unit. This setup allows channels to be selected at the converter-selector.

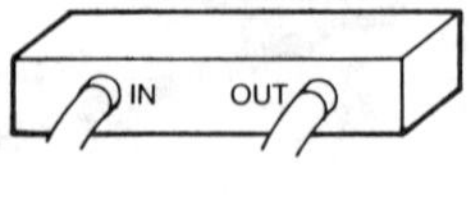

Figure 2-10 Symbol for all types of cable converters used in the hookups.

Converter-Decoder

A converter-decoder is a converter box that also unscrambles subscribed pay TV channels. In the hookups in chapter 6, the converter-decoder always appears as illustrated in figure 2-10.

Pay TV Decoder

A pay TV decoder restores a channel that is purposely distorted by the broadcaster. The distortion is commonly called scrambling. When a scrambled station is tuned without the decoder, the picture can roll, twist, streak, reverse like a photograph negative, or any combination of such effects. The audio can be weak, distorted, or missing. Pay TV decoders are available from a subscription service.

The decoder to play UHF scrambled channels may convert the signal to a VHF channel and switch the VHF antenna through the decoder. In such systems, the VHF antenna signal is passed through for tuning by the TV when the decoder is not being used for unscrambling. When the decoder is turned on, the VHF antenna is disconnected and the decoded-converted signal is output on the VHF leads to the TV. The TV must be tuned to the output channel of the decoder to view the decoded program. Consider the switching function of the decoder box when selecting a hookup.

Videodisc Player

Over the years, a number of videodisc players have been developed. Today these are no longer popular. Special optical videodisc players are still used in industrial and military training applications but rarely used in consumer applications.

The videodisc player, like a VCR, is a signal source. The video program is recorded on a disc that the unit plays to recover the video signal. Most videodisc players contain a built-in RF modulator. The video derived from the videodisc modulates the internal RF modulator, producing a channel 3 or 4 output signal that is connected to your TV set. Usually the output is coax. In some systems, the videodisc player outputs are the video and audio signals themselves, which then can be connected directly to video or audio inputs on the TV set if available. The symbol used to illustrate the videodisc player in the hookups is illustrated in figure 2-11.

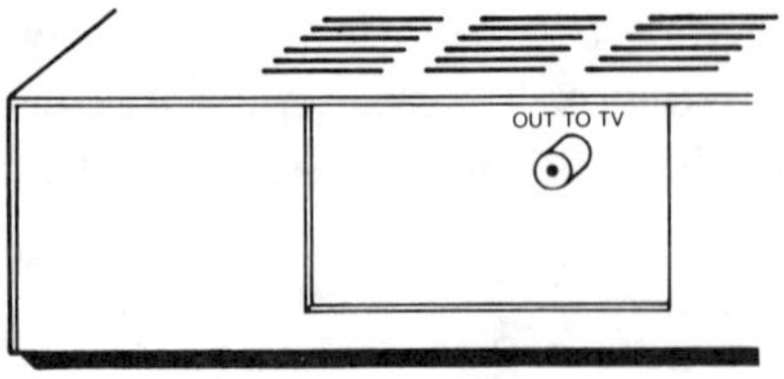

Figure 2-11 Symbol for videodisc player used in the hookups.

Satellite TV Receiver

A satellite TV receiver is a special accessory designed for picking up signals directly from communications satellites. Communications satellites are essentially radio relay stations positioned in space. Most communications satellites are geosynchronous; they revolve around the equator of the earth 22,300 miles away. In this position, they revolve in perfect synchronism with the earth's rotation. Therefore, at any point on earth, they appear fixed in position. Such satellites are used for relaying microwave signals from one point on earth to another. Communications satellites are widely used for long distance telephone conversations and by TV networks, cable TV companies, and others for transmitting TV signals. Although satellites are used primarily by the suppliers of video, consumers can tap this source if they have an appropriate receiver. Because signals are transmitted at extremely high microwave frequencies, special receiving and antenna equipment is required.

The most visible part of any satellite TV receiver is its dish antenna. A large parabolic reflector antenna is required to adequately receive satellite signals. The signals picked up by the antenna are transferred by a large coax cable to the satellite receiver. The receiver itself is simply a special tuner capable of receiving multiple satellite microwave channels. The receiver demodulates the signals into the original video and audio. These signals are then applied to an internal RF modulator that produces a channel 3 or 4 output signal that is connected to the TV receiver or VCR. Some satellite receivers also provide the video and audio signals on the back panel, which can be separately connected to a video monitor or TV receiver with a direct video input. Figure 2-12 shows the symbol used to represent a satellite TV receiver in the hookups.

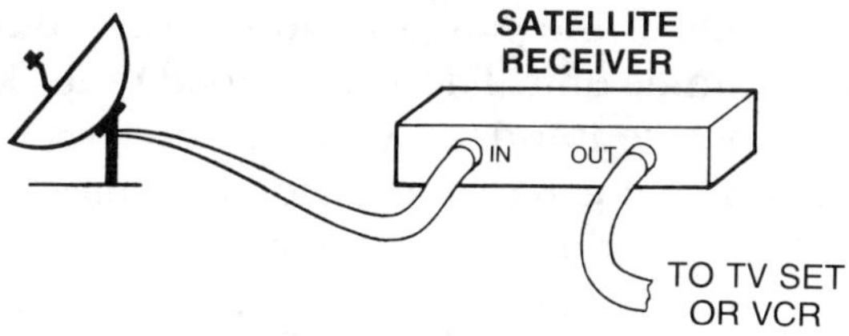

Figure 2-12 Symbol for satellite receiver used in the hookups.

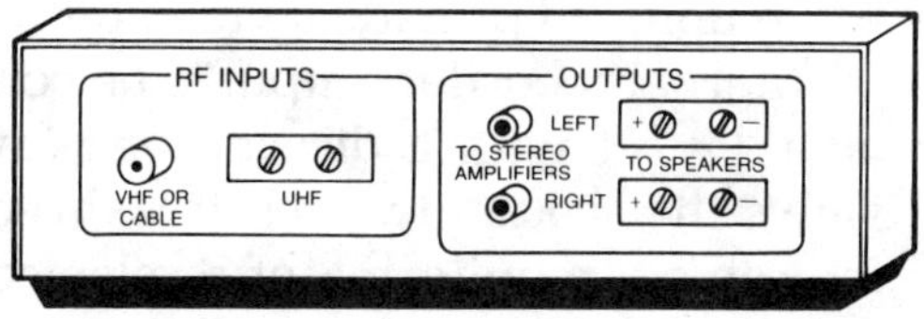

Figure 2-13 Symbol for TV stereo decoder used in the hookups.

TV Stereo Decoder

A TV stereo decoder is a special signal source designed to provide stereo audio signals for older TV sets that do not have a built-in stereo decoder. These units are essentially a TV receiver designed for picking up and demodulating only TV audio signals. The unit receives an input from the usual TV antenna or cable source. The tuner in the stereo decoder is then set to the channel being viewed. If stereo is transmitted, the decoder produces the left and right audio signals. The volume on the standard TV receiver is simply turned down so that the stereo signals provided by the decoder are heard. The outputs of the stereo decoder are usually designed to be fed to a standard stereo hi-fi amplifier system, which contains appropriate power amplifiers and speakers. Other units contain built-in audio amplifiers, which provide low power outputs to accessory speakers. The symbol used to represent a TV stereo decoder is shown in figure 2-13.

Video Camera

With declining prices and numerous technological advances, the video camera has become a popular consumer product. It provides one more signal source for your TV system.

The older video cameras usually worked with a companion VCR. The camera unit contained the vidicon camera tube or the CCD solid-state image sensor and related circuitry. The video signal produced by the camera was connected by a coax cable to the VCR, which recorded the signal on tape. But thanks to miniaturization, the functions of the camera and the recorder are now fully contained within a single unit. These combined units, known as camcorders, are battery operated and fully portable. They do not require a stationary VCR unit. This has made video cameras far more useful and popular.

To "film" an event, a tape is inserted into the camcorder itself. The recorded tape is then played back on a VCR. No coax cable connection between the camera and a recorder or TV set is necessary. However, some units do provide a video output jack, which permits the camera or internal recorder output to be connected directly to a video monitor or a TV set with a direct video input. In this way, the camera just views the scene to which it is directed, and you can use the camera for remote monitoring of a selected area or general surveillance. Or you can use the camera as a playback unit without the VCR.

The symbol in figure 2-14 is used for representing video cameras in the hookups in later chapters. Some cameras use special connectors, so refer to your camera manual for hookup instructions. Usually, special adapter devices or cables are supplied with the camera to simplify hookups.

Headphones

Headphones have been a popular accessory for stereo hi-fi systems for many years. The performance of headphones today is outstanding because the frequency response and the low distortion levels are excellent in modern units.

Headphones are becoming a popular accessory in newer video systems. The problem with using headphones with a TV set is that most TV sets do not provide connections for them. Older TV sets have to be specially modified to accommodate headphones, and these modifications are beyond the ability of the average consumer.

Figure 2-14 Symbol for camcorders or video cameras used in the hookups.

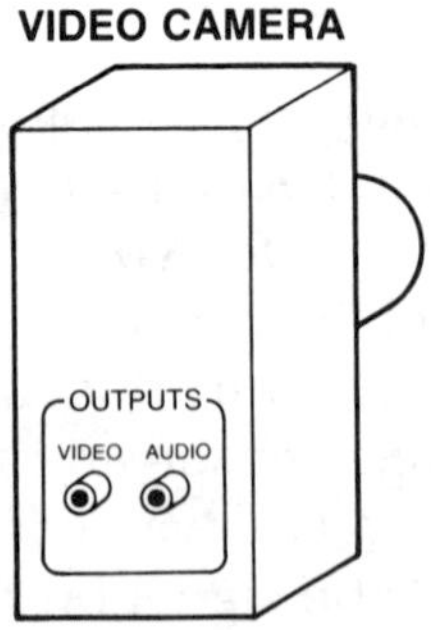

Bookmark

HOWARD W. SAMS & COMPANY

DEAR VALUED CUSTOMER:

Howard W. Sams & Company is dedicated to bringing you timely and authoritative books for your personal and professional library. Our goal is to provide you with excellent technical books written by the most qualified authors. You can assist us in this endeavor by checking the box next to your particular areas of interest.

We appreciate your comments and will use the information to provide you with a more comprehensive selection of titles.

Thank you,

Vice President, Book Publishing
Howard W. Sams & Company

COMPUTER TITLES:

Hardware

☐ Apple 140 ☐ Macintosh I01
☐ Commodore I10
☐ IBM & Compatibles I14

Business Applications

☐ Word Processing J01
☐ Data Base J04
☐ Spreadsheets J02

Operating Systems

☐ MS-DOS K05 ☐ OS/2 K10
☐ CP/M K01 ☐ UNIX K03

Programming Languages

☐ C L03 ☐ Pascal L05
☐ Prolog L12 ☐ Assembly L01
☐ BASIC L02 ☐ HyperTalk L14

Troubleshooting & Repair

☐ Computers S05
☐ Peripherals S10

Other

☐ Communications/Networking M03
☐ AI/Expert Systems T18

ELECTRONICS TITLES:

☐ Amateur Radio T01
☐ Audio T03
☐ Basic Electronics T20
☐ Basic Electricity T21
☐ Electronics Design T12
☐ Electronics Projects T04
☐ Satellites T09

☐ Instrumentation T05
☐ Digital Electronics T11

Troubleshooting & Repair

☐ Audio S11 ☐ Television S04
☐ VCR S01 ☐ Compact Disc S02
☐ Automotive S06
☐ Microwave Oven S03

Other interests or comments: ______________________

____

____

Name ______________________

Title ______________________

Company ______________________

Address ______________________

City ______________________

State/Zip ______________________

Daytime Telephone No. ______________________

A Division of Macmillan, Inc.
4300 West 62nd Street Indianapolis, Indiana 46268

22648

John D. Lenk's Troubleshooting & Repair of Microprocessor-Based Equipment
John D. Lenk
ISBN: 0-672-22476-3, $21.95

Microcomputer Troubleshooting & Repair
John G. Stephenson and Bob Cahill
ISBN: 0-672-22629-4, $21.95

Printer Troubleshooting & Repair
John Heilborn
ISBN: 0-672-22543-3, $19.95

Troubleshooting with the Oscilloscope, Fifth Edition
Robert G. Middleton, Revised by Joseph J. Carr
ISBN: 0-672-22473-9, $16.95

Commodore® 128 Troubleshooting & Repair
John Heilborn
ISBN: 0-672-22554-9, $19.95

John D. Lenk's Troubleshooting & Repair of Audio Equipment
John D. Lenk
ISBN: 0-672-22517-4, $21.95

VCR Troubleshooting & Repair Guide
Robert C. Brenner
ISBN: 0-672-22507-7, $19.95

Place
Postage
Here

HOWARD W. SAMS & COMPANY

Dept. DM
4300 West 62nd Street
Indianapolis, IN 46268-2589

Some newer TV sets have a headphone jack. Plugging in your headphones automatically mutes the speakers for private listening. But the majority of TV sets do not have separate headphone jacks, so special provisions must be made.

One way to obtain headphone operation on a TV set is to connect the TV itself to a hi-fi system and use the headphone jack on the audio amplifier. Most of the newer stereo TV sets have separate audio output jacks in the back that permit connection to a hi-fi stereo system. Simply make these connections (see hookup 20) and use the headphone jack on the stereo system itself.

If your TV set does not have stereo capability, consider purchasing one of the stereo decoder units described previously in this chapter. These units often contain a headphone jack. Even if the stereo decoder does not, it permits connection to an existing hi-fi stereo system that contains headphone capability. Another possibility is to use the stereo audio capability on your VCR if it has it. Many newer VCRs also contain stereo decoders. The audio outputs of the VCR can be used to derive the audio signals, which can drive a separate stereo hi-fi system with headphone capability.

Several companies make a special accessory unit that makes it convenient for connecting headphones to a TV set. The unit is wireless, so it does not require that the headphone cable be connected directly to the TV set. The unit operates on infrared light waves, just like most TV and VCR remote controls.

The wireless headphone accessory has a transmitter unit that connects to the audio output connectors of a TV set or VCR. The audio signals are then used to modulate an infrared ray produced by a light emitting diode. Various methods are used to mute the speakers, including manual switches or remote control. The transmitter unit is mounted on top of the TV set or VCR. It transmits the infrared ray containing the audio to a receiver unit. The receiver unit picks up the infrared signal and demodulates it to provide the original audio for the headphones. Such a system not only gives you listening privacy, but also eliminates the need for long headphone cables.

A Word About Remote Controls

Most modern video equipment is supplied with a handheld remote control unit. Typically, your TV set, VCR, and cable converter box will have remote control capability. If your system contains all three of these devices, you will have to keep all three remote control units handy to operate your system. Often this is a nuisance, but a necessary condition of a sophisticated video system.

If your system contains only a TV and VCR, most of your channel changing will be done with the TV remote control. You will use the VCR remote control for selecting the channel to be recorded.

If you have a cable converter box, channel selection usually is done with the cable box remote control unit. Typically, you will keep your TV and VCR set to channel 3 or 4 and select all other channels with the cable box remote control. The primary functions of the TV and VCR remote controls will be turning power off or on and adjusting volume.

The problem of multiple remote control units has already been addressed by some manufacturers. If you purchase an integrated video/audio system from some manufacturers, an all-in-one remote control unit is supplied. Manufacturers with such systems include RCA and Magnavox. By purchasing all audio and video equipment from the same manufacturer, compatibility is assured and the single remote control will work with all units.

Another approach to solving the remote control problem is to purchase one of the new specialty remote units designed to replace all the remote controls from different manufacturers. A typical unit is called CORE. This unit is capable of learning the functions of all your remotes. You simply program the CORE remote by holding it face-to-face with each of your other remotes. By pressing the keys of the equipment remotes, the CORE remote unit memorizes the specific commands and capabilities. After you have programmed the CORE unit with all of your remotes, it can then take over and control all your units. The CORE remote control unit was designed by Steve Wozniak, one of the inventors of the Apple computer. It is an amazing device and never becomes obsolete because it is always programmable.

It is important to note that the CORE remote is an infrared unit like most other remote controls. However, it is incompatible with the older ultrasonic remote control units used in TV sets built prior to about 1980. Although this unit solves the multiple remote control problem, it is expensive. A typical retail price is about $200.

3 Interconnection Accessories

Interconnection accessories are all those items used to connect the various signal sources to your TV set and to one another. The most obvious accessories are cables and their connectors, but there are many other gadgets as well. These include matching transformers, splitters, switches, and adapters. You will use at least some of these accessories in your hookups.

Where to Get Accessories

Most TV sets, VCRs, and other video units are supplied with several cables and other accessories. When you unpack a new piece of equipment, you will typically find cables, matching transformers, splitters, and other items in the box. Practically everything you need for basic hookups is supplied.

If you want to do more complex hookups, however, most likely you will have to buy extra accessories. These are available from many sources. You may be able to get the accessories you need from the department or electronics store where you bought your equipment. If not, try other stores that sell video and electronic equipment. Many hardware stores also sell TV hookup accessories. You can also get hookup accessories at almost any of the electronic specialty stores, such as Radio Shack and Heath/Zenith Computers and Electronics.

Connecting Cables

Connecting cables are the most important hookup accessory you will use. Cables are the wires that get the signals from one piece of equipment to another. In general, a cable is nothing more than two or more wire conductors especially made for making electrical connections. There are two basic kinds of interconnecting cables used in video systems: the antenna or signal source cables that carry RF signals and the interconnecting cables for video and audio signals. The antenna or signal source cables carry the signals from your antenna or cable system to the equipment. The interconnecting cables connect your video units together.

Twin Lead

There are two types of cables used in video hookups, twin lead and coaxial cable. These are referred to as transmission lines. Figure 3-1A shows the details of a typical twin lead cable. The two stranded copper wires are separated by a continuous plastic insulator about ½-inch wide. These wires can be connected directly to the screw terminals on the equipment. Often twin lead is supplied with special connectors called spade lugs, which are attached to the wires. Figure 3-1B shows twin lead cable with special connecting lugs on each end.

Coaxial

Coaxial (coax) cable is the other kind of cable widely used in video connections. It is a round cable with a wire center conductor surrounded by a wire braid. Figure 3-2A shows the construction of a typical coax cable. Plastic or teflon insulation surrounds the solid center conductor. A wire braid surrounds the insulation. Finally, an in-

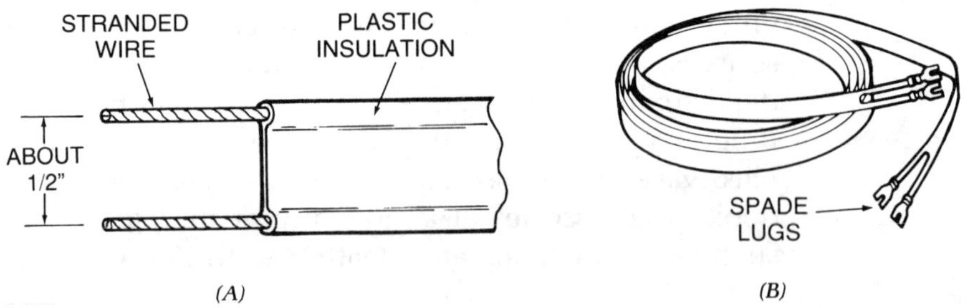

Figure 3-1 Twin lead transmission line.

Figure 3-2
Coaxial cable.

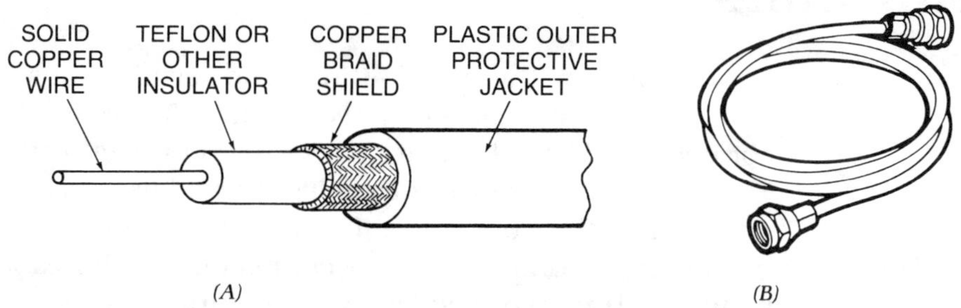

sulating outer sheath of plastic protects the braid. Figure 3-2B shows a coil of coax with special connectors on each end.

Coax cable is made in a variety of sizes. The size most commonly used in video interconnections has the designation RG-59/U. It is approximately ¼ inch in diameter. Coax cable can be purchased with connectors already installed in lengths of 2, 4, 6, 8, and 10 feet.

Both twin lead and coax cables, as well as the RF and video connections on the equipment, have an electrical property known as impedance. You do not need to know what impedance is, but you do need to know how it can affect your hookup. Throughout any hookup, you have to be sure that impedances always match.

Twin lead has an impedance of 300 ohms. Most TV antennas have a 300-ohm impedance. The VHF and UHF antenna terminals on TV sets, VCRs, and other units have a 300-ohm impedance.

Coax cable has an impedance of 75 ohms. The VHF antenna or cable input on a TV set or VCR is 75 ohms. All coax video cables and connectors have an impedance of 75 ohms.

You cannot connect a 300-ohm piece of equipment to a 75-ohm cable or vice versa. Not only would the physical interconnections be difficult, but also the resulting mismatch would produce tremendous signal attenuation or distortion. Picture quality would degrade. Many times you will find that a signal source has a 300-ohm impedance but the unit to which it is to be connected has a 75-ohm input impedance. Sometimes the signal source has a 75-ohm impedance but the input impedance is 300 ohms. In such cases, the impedances must be matched for proper operation. This is accomplished with a unit called a matching transformer (described later).

There is one final point to keep in mind. Cables have attenuation. That is, they introduce losses that reduce signal level. Because this attenuation is proportional to cable length, keep cables short for best results.

Connectors

Connectors are the metal fittings and terminations used for conveniently connecting cables and equipment. There are two basic kinds of connectors. First, there are those that connect to the ends of cables. The wire conductors in the cables are soldered, crimped, or otherwise attached to the connector. The other type of connector is mounted on the video unit itself. The connector on the cable is designed to match with the connector on the equipment.

Twin Lead Connectors

The connectors used on twin lead are extremely simple. The most common connector is a solder lug or spade lug, as illustrated in figure 3-3. These are small metal pieces crimped or soldered to the wires at the ends of the cable. The lugs have a notch at the end that make them easy to slip under a screw.

Cables supplied with video equipment or accessory kits purchased separately already have spade lugs attached. If not, you can purchase and install the spade lugs. In any case, the use of spade lugs speeds up and simplifies interconnections.

In most cases, it is not even necessary to use spade lugs on the twin lead. Using a knife, you can strip the insulation from the twin lead to expose the stranded wires. This is shown in figure 3-4. The

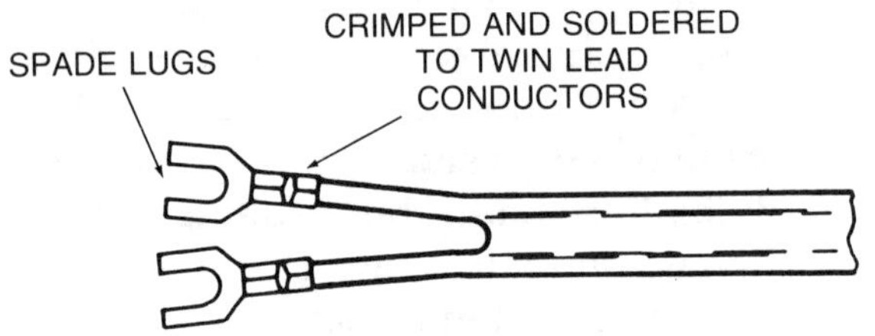

Figure 3-3 Spade lug terminals for 300-ohm twin lead.

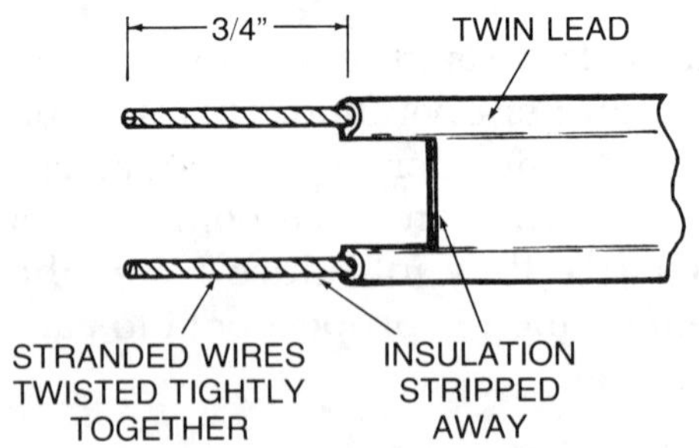

Figure 3-4 Stripping 300-ohm twin lead.

Figure 3-5 Installing twin lead on screw terminals.

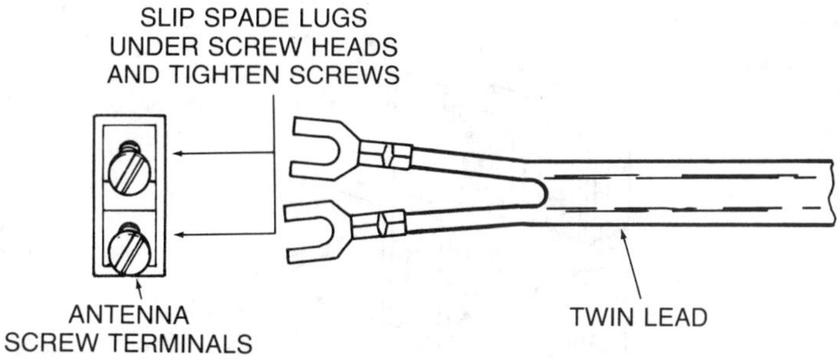

stranded wires can then be twisted tightly together, then wrapped around the screw terminals.

The 300-ohm connectors on video equipment and accessories are typically screw terminals. Usually, two 6-32 flat head screws are used. The spade lugs are slipped under the screw heads, and the screws are tightened to make the connection. See figure 3-5. Alternatively, the twin lead wires can be wrapped around the screw terminal and tightened. If the wires are used, be sure to wrap the wire around the screw in a clockwise direction as illustrated in figure 3-6A. In this way, the wire will be pulled in the proper direction for making a good connection. If the wire is wrapped around in a counterclockwise direction, when the screw is tightened the wire will have a tendency to unwrap and a poor connection can result. See figure 3-6B.

Another type of connector becoming popular for twin lead connection is the push-in pressure terminal. A typical unit is shown in figure 3-7. These are popular on audio equipment for connecting speakers. To use this connector, you first strip the twin lead as described previously. Then you press on the lever and insert the bare wire fully into the hole. Then release the lever. A spring tension metal

Figure 3-6 Attaching twin lead wire to screw terminals.

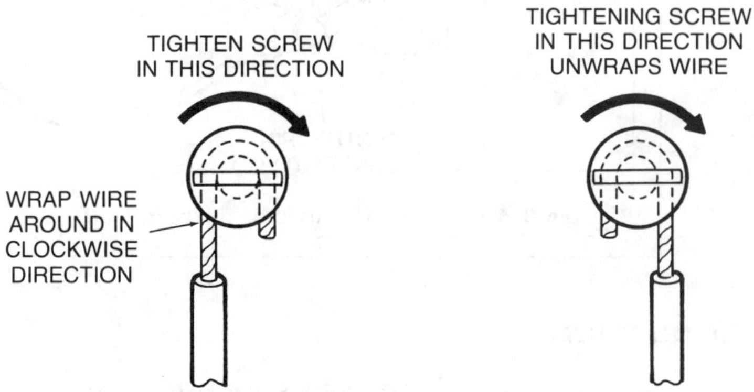

(A) Right *(B) Wrong*

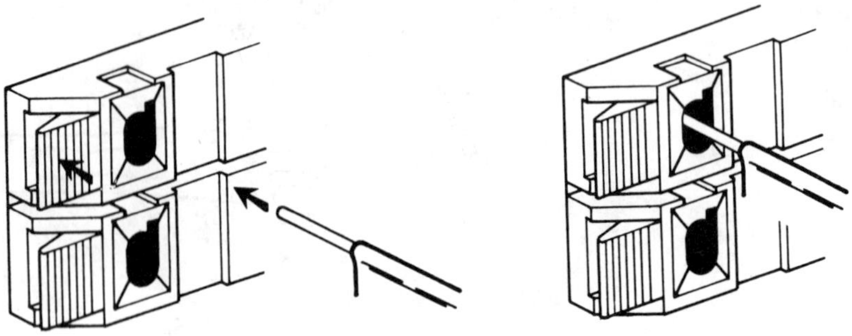

Figure 3-7 Push-in pressure terminal for 300-ohm twin lead and speaker wire.

contact inside holds the wire so that it is mechanically secure and makes a good electrical connection.

Clothespin Clip

If you use a lot of twin lead in your hookups, you may want to take advantage of a special connector known as a clothespin clip. See figure 3-A. The twin lead attaches to screw terminals on the clip. The clip is then attached to the screw terminals on the TV set or VCR. Squeezing the clip separates the metal connections and allows them to be clipped to the screw terminals. The screws should be unscrewed about ⅛ inch so there is space for the clip. Spring tension keeps the clip in place and helps make a good electrical connection. Such a clip makes connections faster and easier and allows more convenient hookups.

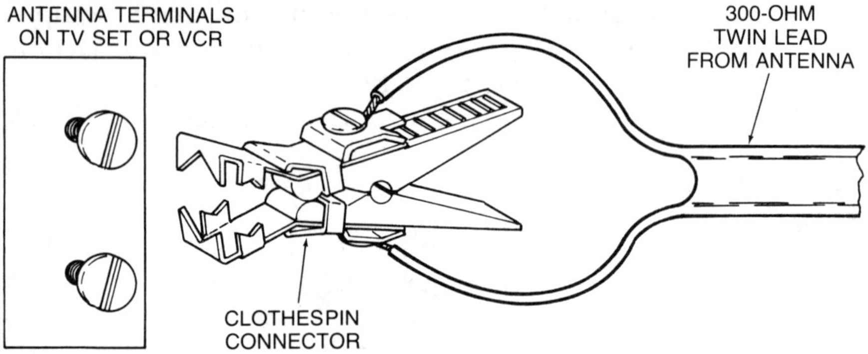

Figure 3-A Clothespin connector for twin lead.

F-Connectors

Several types of connectors are used with coax cable. The most common types are F-connectors, phono connectors, and miniconnectors.

An F-connector used on the end of a coax cable is illustrated in figure 3-8. It consists of an outer metal shield attached to the braid of the coax. This outer metal shield can be rotated and contains threads on the inside so that it can be screwed onto a matching connector on the equipment. The stiff center wire conductor of the coax cable forms the center connection of an F-connector.

Figure 3-8
F-connector.

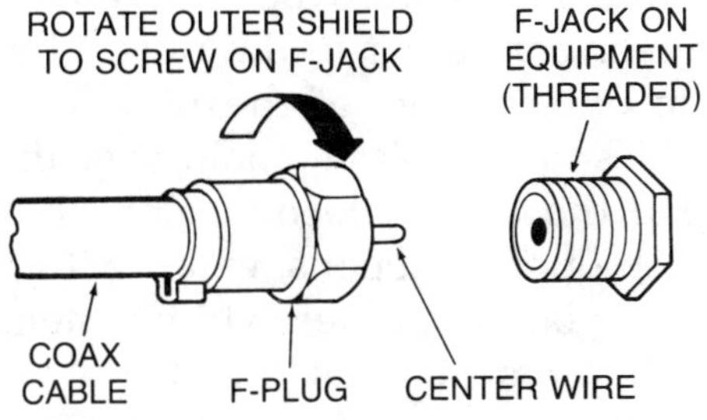

F-connectors attached to the ends of cables are usually referred to as male connectors or plugs. On the video equipment to be interconnected are matching F-connectors referred to as female connectors or jacks. The center conductor fits into the center hole of the F-jack and the outer sleeve is screwed onto the threads of the jack. The result is a reliable coax connection with the proper impedance.

A popular version of an F-connector plug is the push-on or slip-on type frequently supplied on the cables packed with VCRs. See figure 3-9. The center conductor of the cable is still used, but a smooth cylindrical outer sleeve is used instead of a threaded outer sleeve. The smooth outer sleeve simply pushes on over the threads on the F-jack. The slip-on cable is much faster and easier to interconnect than the screw-on type. Because the fitting is tight, a good electrical connection is made.

Figure 3-9
Slip-on F-connectors.

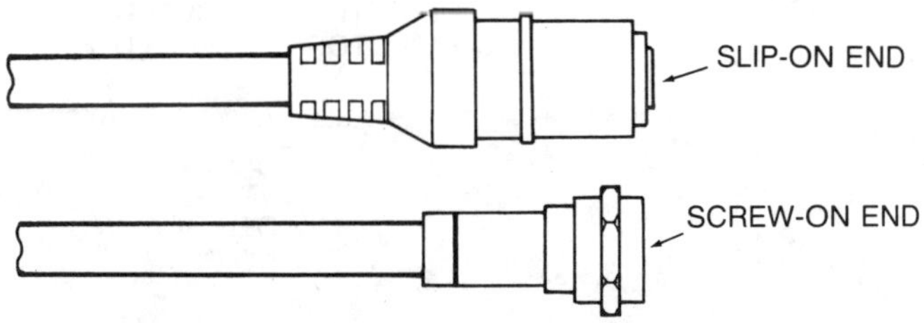

RCA Phono Connectors

Another type of connector widely used with coax is the phono connector. These connectors were originally developed by RCA for interconnecting phonographs and other audio equipment. However, because they were designed for coaxial cable, they also make good connectors for video signals.

Figure 3-10 shows RCA phono plugs and jack. The plugs are usually connected to the cable ends and the jack is used on the equipment. The plugs are soldered to the cable ends. Unlike F-connectors, phono connectors require the coax center conductor to be soldered to a larger center conductor in the plug. The coax braided shield is soldered to the outer sleeve of the phono connector. Most commercially built cables have a plastic mold over the connector and cable for strength. Like the slip-on F-connector, the phono connector is designed for push-on connections, making it fast and easy to use.

If you have a stereo hi-fi system, you are no doubt already familiar with phono connectors. These are the kind of connectors used to attach stereo receivers, amplifiers, cassette players, CD players, and other audio units to one another. Many standard cables with connectors are readily available.

Phono connectors are common on cables used for interconnecting various pieces of equipment. Coax cables with phono connectors are used primarily to carry audio and video signals. F-connectors are used in carrying both VHF and UHF RF signals.

Although connectors can be purchased separately for installation on coax cable, it is best to buy cables with connectors already attached. This will save you time and the frustration of connector installation. Different grades of connectors are available. The best ones are gold plated. Gold is an excellent electrical conductor that does not easily oxidize or corrode. Gold-plated connectors also eliminate noise but are more expensive than standard connectors.

Figure 3-10 RCA phono connectors are used for video and audio interconnections.

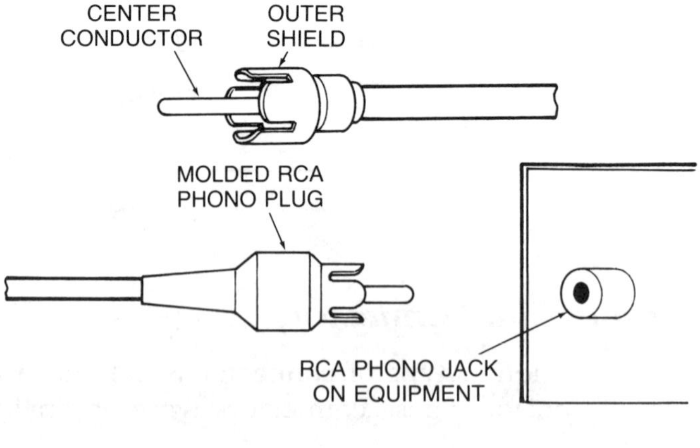

Miniconnectors

Miniconnectors are used mainly for audio connections. They are widely used for headphones and, as a result, are sometimes referred to as phone plugs. They are also used for video in some systems. Beta VCRs use them for video inputs and outputs.

Figure 3-11 shows a miniplug and minijack. The plug is used on the ends of coax or shielded wire. The center conductor is soldered to the plug tip and the braid is soldered to the barrel. The barrel is typically 3.5 mm in diameter. The plastic sleeve screws off, giving access to the connections.

Figure 3-11
Miniconnectors.

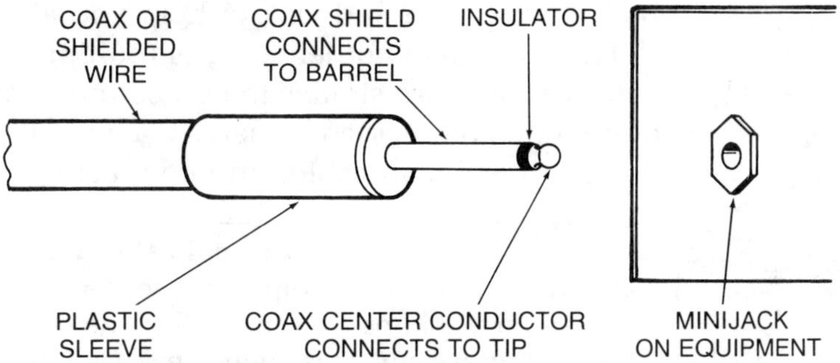

Miniplugs used on stereo headphones have three connections. They are used with a special two-conductor shielded cable. The shield is connected to the barrel, and the two wires inside the shield connect to the tip and a ring. See figure 3-12. The insulators keep the connections isolated. Some headphones use a three-conductor unshielded cable.

Figure 3-12
Miniplug used on stereo headphones.

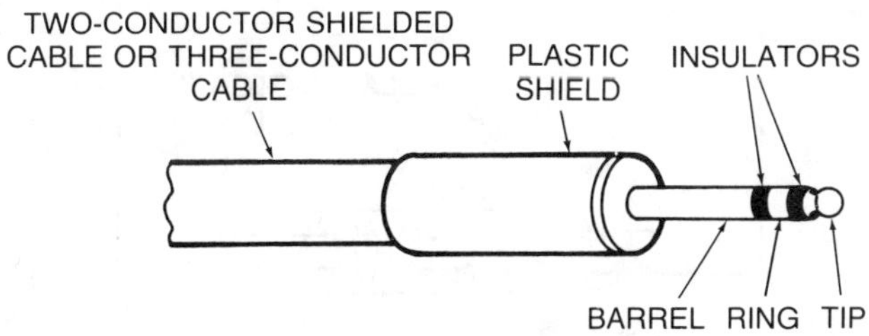

Multipin Connectors

To minimize the number of cables and connectors required in a system, some manufacturers use multiconductor (more than two) cables and multipin connectors. These are used for video and audio signals, not RF.

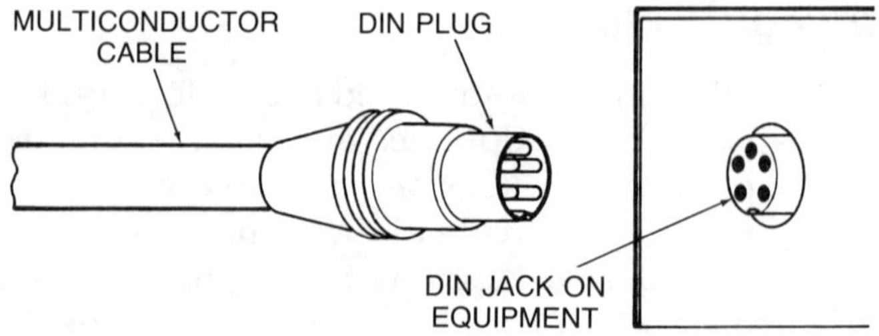

Figure 3-13 DIN connectors.

Figure 3-13 shows one common type of multipin connector known as a DIN connector. DIN connectors have 3 to 8 pins, but the 3- and 5-pin versions are the most common. DIN connectors have a key or notch on the plug and jack to help you align the two.

There are many other types of multipin connectors. XLR-type connectors, which are similar to DIN connectors, are used in some audio equipment. Some video cameras use miniature connectors with 10 or 14 pins. Luckily, all cables and connectors are usually supplied.

Motorola Connector

Although most video equipment uses F-connectors or RCA phono connectors for hookups, you might find the connector illustrated in figure 3-B in your system. This connector, made by Motorola, is widely used in some cable TV or master antenna systems in apartments and condos. A special jack on the wall terminates the coax transmission line containing the RF signals. The plug in figure 3-B is inserted into the jack to transfer the signal to the TV set or VCR. A special cable or adapter may be needed if you have such a connector.

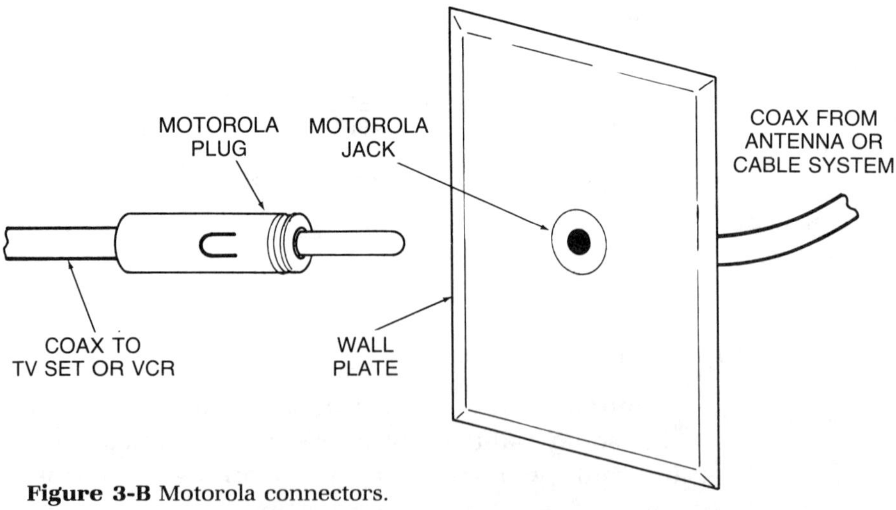

Figure 3-B Motorola connectors.

Adapter Jacks and Splicers

Adapter jacks are special connectors designed to join mismatched sets of connectors and cables. They are used primarily on coax interconnections. Figure 3-14 shows two of the many types of adapters available. The ones you are most likely to use are phono jack to F-plug or phono plug to F-jack.

Figure 3-14 Adapter jacks.

If you are making audio interconnections, you may be using adapters for phono connectors as well as the various types of miniplugs and minijacks often used on stereo equipment. Miniplug to phono jack is another common arrangement.

A splicer is a special connector used to join two cables or a cable and an accessory. A typical F-connector barrel splicer is shown in figure 3-15A. This device has female F-connectors on each end. Cables containing male F-connectors can be conveniently attached using this device while maintaining the proper 75-ohm impedance match. This eliminates the need to splice cables directly by soldering.

Another such connector is shown in figure 3-15B. It features screw terminals on either end for 300-ohm twin lead. Such a device, although not as widely used as the barrel splicer, is handy in some hookups.

There are dozens of different types of adapters and splicers. We can't cover all of them in this book. As you are performing your hookup, you may need a special adapter. Check your local electronics store to see if the kind of adapter you need is available.

Figure 3-15 Splicers.

(A) 75-ohm F-barrel splicer *(B) 300-ohm 4-screw block*

Matching Transformers

The connectors on the back of your TV set, VCR, and other video equipment are either inputs or outputs. Some connectors accept input signals from a signal source. Output connectors supply signals to some other piece of equipment. These input and output connections have specific impedances, just as the cables described previously. If screw terminals are used, the input or output impedance is 300 ohms. If coax connectors are used, either F-type or phono type, the input or output impedance is 75 ohms. Usually, the proper cable and connector will allow you to make the correct interconnection without difficulty.

Many times it is necessary or desirable to connect a 300-ohm source to a 75-ohm input or vice versa. The connection usually would be difficult due to the wide differences in connector types, and the resulting impedance mismatch would cause problems such as signal attenuation, distortion, and improper operation.

To take care of these situations, a special accessory known as a matching transformer is used. A matching transformer (also called a balun) allows the connection of a 300-ohm source to a 75-ohm input or a 75-ohm source to a 300-ohm input.

Figure 3-16 shows two types of matching transformers. The unit at A features screw terminals for the 300-ohm twin lead and a right-angle slip-on F-plug. The type at B features an F-jack and a short piece of 300-ohm twin lead with spade lugs.

Matching transformers are usually bidirectional. This means either of the two connections can be used as the input or the output. For example, the connector in figure 3-16A could be used for attaching the twin lead from a 300-ohm antenna to the 75-ohm input to a VCR. On the other hand, the connector in figure 3-16B could be used to connect the 75-ohm VHF output from a VCR to the 300-ohm antenna terminals on your TV set. It doesn't matter which of the connections are used as inputs or outputs because the impedances match.

There is one important point to keep in mind when using matching transformers. These devices usually introduce some signal attenuation. That is, some of the signal is lost in the impedance transformation process. If the signal amplitudes are high enough, you will not notice the small amount of attenuation. But in applications where signal strength is marginal, matching transformers can make it worse. Your

Figure 3-16 Matching transformers.

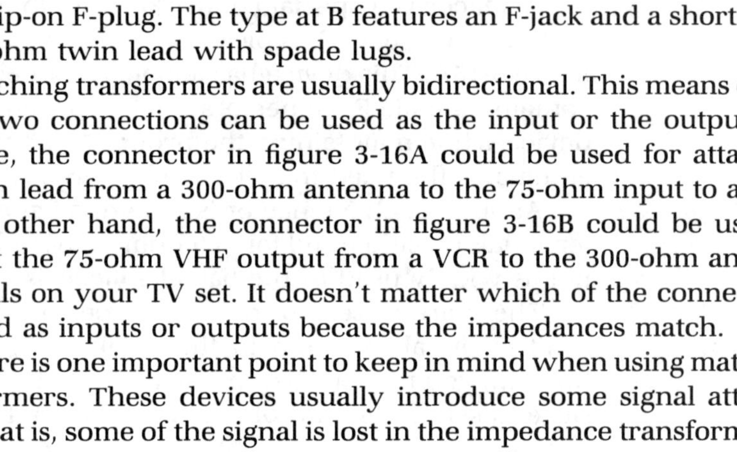

(A) 300-ohm screw terminals to 75-ohm right-angle F-plug

(B) 75-ohm in-line F-plug to short 300-ohm twin lead with spade lugs

picture may be snowy as a result. In some applications, several matching transformers are necessary to achieve the desired hookup. The more matching transformers you use, the more loss you introduce. Try to minimize the use of these devices where possible for the best possible signal performance.

Band Separators

A band separator, also known as a band splitter, is a device used for separating RF signals in different parts of the frequency spectrum. More specifically, the band separator takes RF signals such as the VHF and UHF TV signals that have been combined at some point and separates them into different outputs that can be connected to the VHF and UHF inputs on TV sets and VCRs.

One reason a separator is needed is that TV signals are often combined at some point in the system. A very common occurrence is the use of an antenna that receives both VHF and UHF signals. These two bands of signals are transferred from the antenna to the TV set by way of a single 300-ohm twin lead or coax cable. At the TV set or VCR, separate VHF and UHF inputs are typical. The function of the band separator is to take the combined signals from the single transmission line and separate them into different VHF and UHF outputs that connect to the TV set.

Some TV antennas are broadband and, therefore, capable of receiving signals over a very wide frequency range. In fact, in addition to VHF and UHF TV signals, FM radio signals are received. These three are combined on the single transmission line. Special band separators are available with three outputs for separating a single composite signal into VHF, UHF, and FM outputs.

Band separators have two important functions. The first and most important is frequency discrimination. Band separators contain filter circuits made of inductors and capacitors that are designed to select a particular band of frequencies while rejecting others. These filters are connected to the common input and their outputs are sent to separate leads. Figure 3-17 shows the functional block diagram of common band separators.

Another important function of the band separator is impedance matching. In addition to providing frequency discrimination, each separator must ensure that the impedances of all inputs and outputs are properly matched. Because of the variety of different transmission

Figure 3-17
Band separators.

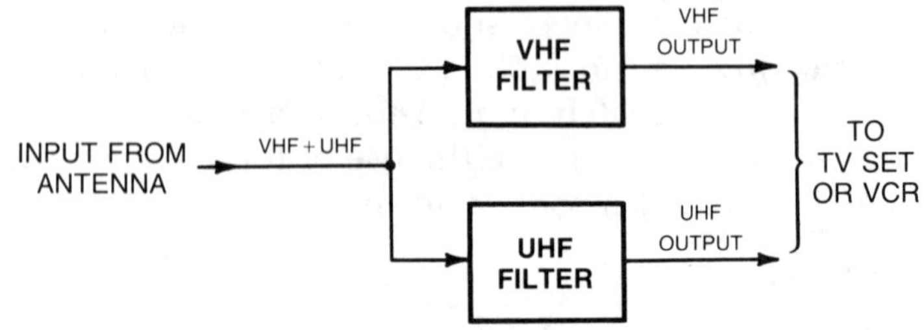

(A) VHF and UHF

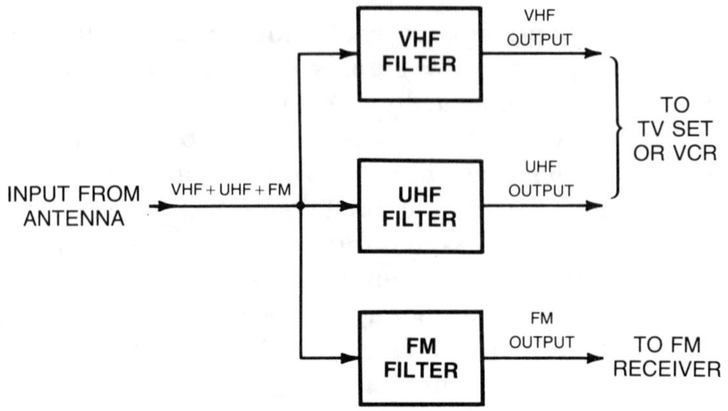

(B) VHF, UHF, and FM

lines and input and output impedances of TV sets and VCRs, many different band separators are available. For example, band separators can have a single 300-ohm twin lead input or a coaxial input. Outputs can be 75-ohm coax, 300-ohm twin lead, or some combination of the two. Because of this impedance matching function, band separators are sometimes referred to as band splitter-matching couplers. Figure 3-18 shows a variety of band separators. Each has a single input and two outputs. Figure 3-19 shows a unit with a 300-ohm antenna input and FM, VHF, and UHF 300-ohm outputs. If you are using a TV antenna as your main signal source and you can receive both UHF and VHF signals in your area, you will undoubtedly need a band separator. Often such a unit is supplied with the TV antenna. If not, you can purchase one separately. Select the one with the arrangement of input and output impedances to suit your situation.

One final precaution about using band separators. Like matching transformers, band separators introduce attenuation. The circuits provide filtering and impedance matching but absorb some of the signal in the process, making the output level lower than the input level.

Figure 3-18 Band splitter-matching couplers.

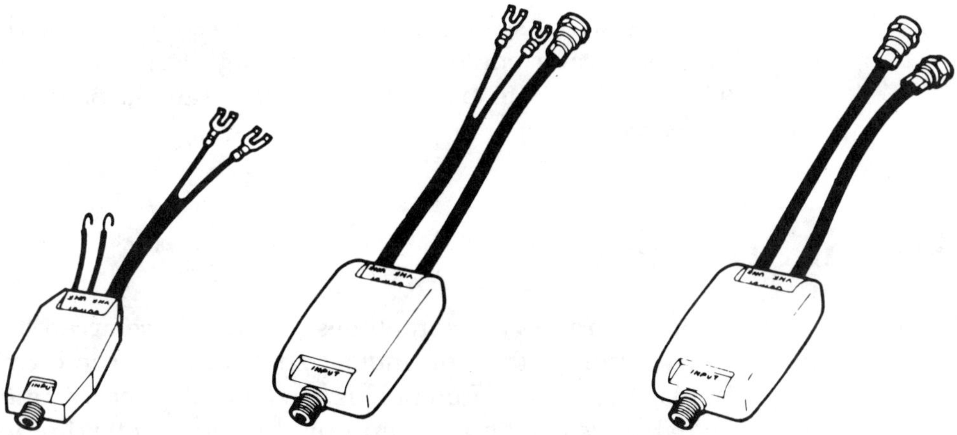

(A) 75-ohm input, two 300-ohm outputs

(B) 75-ohm input, one 300-ohm output, one 75-ohm output

(C) 75-ohm input, two 75-ohm outputs

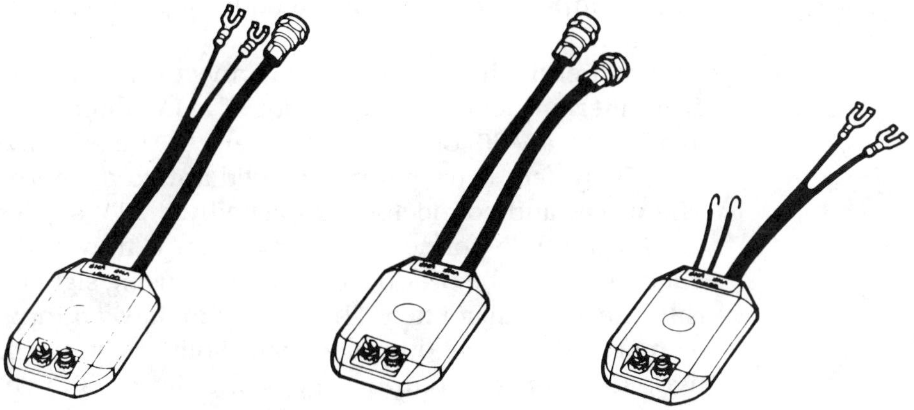

(D) 300-ohm input, one 300-ohm output, one 75-ohm output

(E) 300-ohm input, two 75-ohm outputs

(F) 300-ohm input, two 300-ohm outputs

Figure 3-19 Band splitter with VHF, UHF, and FM outputs.

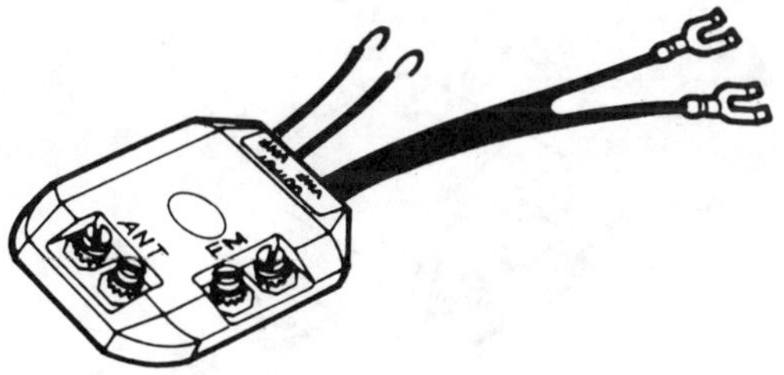

Because signal strength is reduced, in some marginal signal areas a band separator could introduce enough loss to cause poor reception. The solution is to use a higher gain antenna or install an amplifier (discussed later).

Signal Splitters

Band separators are sometimes mistakenly referred to as signal splitters. Although they do indeed split the signal into different components, their main function is frequency discrimination rather than signal redistribution. When purchasing or referring to a device that separates different frequencies, use the term *band separator*. The term *band splitter* is acceptable (and is often seen in catalogs), but in general, the term *splitter* should be reserved for referring to devices that divide a single input into two or more equal signal levels without frequency selection.

Frequently, it is desirable to connect the same input signal to several different devices. The output of a TV antenna can be connected to a TV set, a VCR, or some other device. The output of the cable from a CATV system is frequently needed at two or more different inputs. The cables and connectors associated with TV signals do not actually permit multiple connections. And even if they did, impedance mismatches would be produced, thereby causing signal loss. The solution is to use a signal splitter. The splitter divides an input signal into two or more equal level output signals. Splitters are available to produce two, three, or four outputs. Typical units are shown in figure 3-20.

In addition to providing multiple versions of the same signal, the splitter performs all necessary impedance matching functions. Splitters are designed only to work with coax cable and will maintain constant 75-ohm input and output impedances. Sometimes signal splitters are referred to as multiset couplers. Some manufacturers refer to their splitters as two-, three-, or four-set couplers because they permit

Figure 3-20
Signal splitters.

(A) 75-ohm one-to-two

(B) 75-ohm one-to-four

two or more TV sets, VCRs, or other units to share a single signal source.

Like matching transformers and band separators, signal splitters also introduce attenuation. To provide an equal division of an input signal while maintaining a constant impedance, signal splitters must use resistor networks to divide the signal. These resistors introduce insertion loss. Because this attenuation is significant, these devices work best where a strong input signal is available. When the signal from the TV antenna or cable system is strong, the splitter loss will not be noticed. But in marginal cases, the attenuation introduced by the splitter can cause unacceptable reception. If possible, signal splitters should be avoided to eliminate this problem. If an application requires splitters because a single signal source must be shared among different pieces of equipment, an amplifier may be required to offset the splitter insertion loss.

The input/output jacks on the VCR can act as a splitter. For example, to split an antenna signal to a VCR and a TV set, you can connect the antenna to the VCR terminals and connect the VCR RF output to the TV antenna terminals.

Terminators

A terminator is a special coax fitting used with signal splitters to ensure proper impedance matching on unused outputs. For example, suppose only three outputs on a 4-way splitter are used. If the fourth output is not connected to another cable of the correct 75-ohm impedance, an impedance mismatch will occur. This upsets the operation of the entire device, introducing unwanted signal attenuation and distortion.

To overcome this problem, a terminator is used. This is a special type of coax fitting that screws onto the unused output of the splitter. The terminator contains a 75-ohm resistor that provides the correct impedance match. A typical terminator is shown in figure 3-21. Although such terminators are used primarily with signal splitters, they can be used in any application where it is necessary to terminate an unused output on other devices.

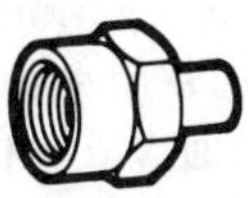

Figure 3-21
Terminator plug.

Attenuators

An attenuator is an accessory specifically designed to introduce a certain amount of attenuation or loss to a TV signal. In many cases, TV signals are barely strong enough to provide adequate performance. But sometimes excessive signal strength is available. If your home is located near TV transmitting stations, high signal strength at your receiver can cause severe overloading, which results in distortion of both picture and sound. The only solution is to reduce the signal strength.

Excessive signal strength can also come from an antenna with very high gain. Large, multi-element, highly directional TV antennas provide a tremendous amount of gain that simply may not be necessary in an installation. The result is an extremely high signal level on the transmission line. It too can provide signal overload and distortion.

Some TV sets and tuners also contain extremely sensitive tuners that do not require high levels of signals for outstanding performance. Such tuners are more easily overloaded by strong signals.

To correct this, you can use an attenuator. Attenuators are designed for installation in 75-ohm coax lines and will greatly reduce the signal level. Some 300-ohm attenuators are also available. If your system contains a number of splitters, separators, matching transformers, and other coupling devices, their combined loss may be sufficient to reduce the signal strength and eliminate the need for a separate attenuator.

Several types of attenuators are shown in figure 3-22. Fixed attenuators introduce a specific amount of loss. The usual attenuator has F-jacks on both input and output so that it can be connected easily to the system with coaxial cables and F-plugs. Typically, the attenuator is connected in the transmission line directly from the TV antenna before the signal is distributed to other devices and equipment. A small, cylindrical, fixed in-line attenuator is illustrated in figure 3-22A.

Although most attenuators introduce a fixed amount of attenuation, variable attenuators are also available. These contain a built-in rotary control that allows you to adjust the attenuation to optimize the signal strength for your installation. Both 75-ohm coax and 300-ohm twin lead models are shown in figure 3-22B.

Attenuation or loss is usually expressed in units called decibels, abbreviated dB. (Decibels are explained in a later section.) Fixed in-line attenuators are available in 3, 6 and 10 dB loss models. Variable attenuators can vary the loss from 0 to 15 dB.

Figure 3-22
Attenuators.

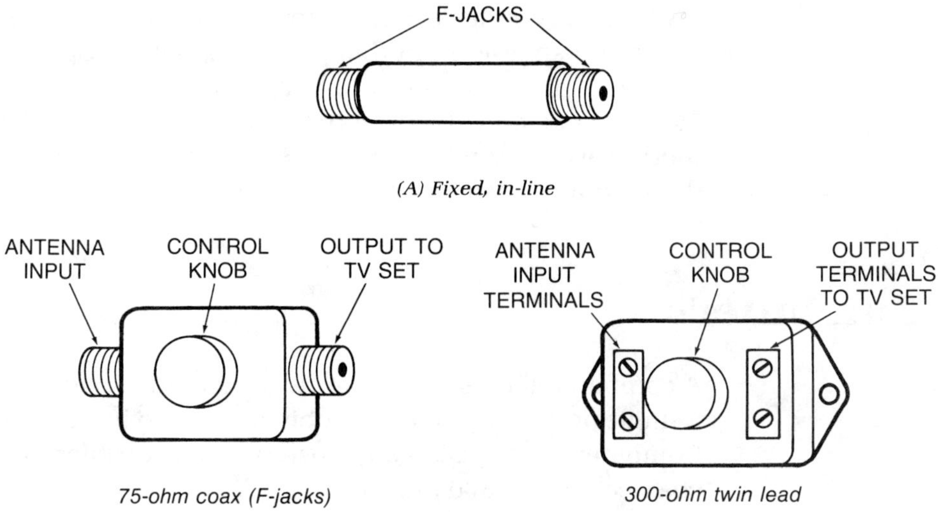

(B) Variable

Interference Filters

A filter is a frequency selective device that allows some frequencies to pass with minimum attenuation and eliminates or greatly attenuates other frequencies. The band separators discussed previously are a special type of filter used for separating a composite signal into multiple outputs on different frequency bands. Filters can also be used for eliminating interference from unwanted sources. For example, transmissions from amateur (ham) radio transmitters and citizens band (CB) radio equipment can produce television interference (TVI). Most amateur radio and CB equipment operate on frequencies lower than TV stations. However, these signals produce harmonics, or multiples of the transmission frequencies, that often fall into the TV bands. Despite the low strength of these signals, they often produce interference to both picture and sound.

To eliminate this problem, special interference filters are available. These devices are known as high-pass filters because they allow the high-frequency VHF and UHF TV signals to pass while they filter out the lower frequency amateur radio and citizens band signals.

Another special form of interference filter is called a trap. A trap eliminates an interfering signal on a specific frequency or a narrow band of frequencies. It is often called a band stop or notch filter. A typical trap is one that eliminates TV interference in the form of a herringbone pattern caused by strong FM radio stations.

Typically, interference filters are connected directly into the antenna transmission line before the signal is applied to the TV set or VCR and prior to any couplers, splitters, or other accessories. Both 75-ohm coax and 300-ohm twin lead filters are available. Some typical models are shown in figure 3-23. Like other devices of this kind, filters do introduce a small amount of attenuation.

Combiners

A combiner allows two separate signal sources to be combined to form a single output. A typical combiner has two inputs and a single output. Combiners are available for use with coax cables or a combination of coax cable and 300-ohm twin lead.

A typical combiner looks very much like a 2-output signal splitter. In fact, many signal splitters are bidirectional and can be used as combiners as well as splitters. Most signal splitters are bidirectional or reversible because they are made of resistor networks. Outputs can be used as inputs and vice versa. Often a signal splitter can be used as a combiner as a result. Special combiner units are also available.

Combiners are frequently needed in applications where more than one signal source exists. In some installations, separate VHF and UHF antennas are used. It is usually convenient to combine the outputs

Figure 3-23 High-pass interference filters.

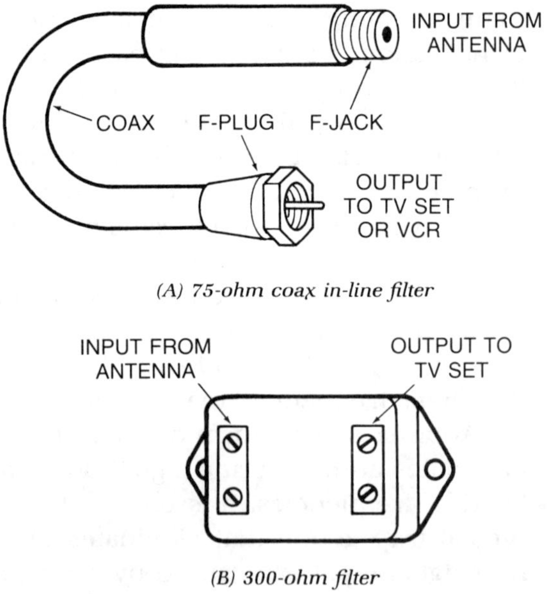

(A) 75-ohm coax in-line filter

(B) 300-ohm filter

from these antennas so that only a single transmission line is needed. The combiner is usually mounted outside near the antennas, and the signals are brought in through a single coax cable.

RF Modulators

An RF modulator is a miniature radio transmitter designed to produce an output signal that can be applied to the TV set antenna terminals. It can take a video input signal from some source and add it to the carrier signal on channel 3 or 4. The RF modulator allows video signals to be displayed on a TV set when other means are not provided.

Most RF modulators are used with TV games, personal computers, and other devices that produce a video output signal rather than an RF output signal in the VHF or UHF TV bands. If you are using one of the newer TV sets or monitors, a video input connector is available on the back of the set and, therefore, the video output from the game, computer, or other device can be attached directly. But if no video input connector is available, how can the video be displayed? The answer is to apply it to an RF modulator. Some video games have a built-in RF modulator.

The RF modulator consists of an electronic circuit known as an oscillator that generates a TV signal on either channel 3 or channel 4. Some units have been designed for UHF output. Usually a switch is provided to select the desired channel. The RF modulator unit also contains another circuit known as a modulator. The modulator modifies the output of the oscillator in a special way, causing it to impress the video input onto the carrier signal generated by the oscillator. RF modulators perform amplitude modulation (AM) of the carrier. That is, the video signal varies the amplitude of the carrier output in accordance with the video information. The result is a TV signal exactly like that received by the antenna.

Some RF modulators also have an audio input jack. The audio signal frequency modulates another RF oscillator on another frequency. The outputs of both carriers are combined into a single RF output signal.

Although most RF modulators are transmitters, they are not designed to be used with antennas. Instead, the signal output is connected directly to the antenna terminals of the TV set. Some RF modulators have a 300-ohm output impedance. Others have a 75-ohm output. The video and audio inputs to the RF modulator are typically

RCA phono jacks. The RF output is either 300-ohm screw terminals or a 75-ohm coax F-jack.

Because an RF modulator contains electronic circuitry, some means must be used to provide power to it. Usually, an internal power supply circuit operates from a separate AC line cord or transformer that plugs into a wall outlet.

Most RF modulators are not necessary in modern installations simply because these units are often built into other equipment. For example, every VCR contains an RF modulator. The video signal available at the output of the VCR tuner or derived from a tape it is playing is fed to an internal RF modulator that produces the RF signal connected to the TV set for display. Satellite TV receivers and other accessories also often contain an RF modulator.

Amplifiers

An amplifier is an electronic circuit that increases the signal level. A weak signal applied to the input will be greatly amplified. Amplifiers are used to overcome the loss and attenuation introduced by using many of the accessories described previously. They are also used in weak signal areas where the inherent amplification of the TV set, VCR, or other equipment is simply inadequate. Amplifiers are also used when you want to distribute a video signal to two or more different accessories through splitters.

Amplifiers are available in numerous configurations. They are available for use with either 75-ohm or 300-ohm input and output impedances. They are also available in versions with multiple outputs. This allows the signal to be distributed to several sets or accessories, thereby eliminating the need for a separate signal splitter.

The simplest form of amplifier has a single input and single output. Such a simple amplifier is usually extremely small and can be easily connected and hidden behind the. These small amplifiers are often referred to as in-line amplifiers because they are approximately the same size as a matching transformer or band separator. See figure 3-24A.

Larger, more complex amplifiers are also available. One type is designed to mount outside at the TV antenna. The signal from the antenna is connected with a short piece of 300-ohm twin lead or coax to the amplifier. The amplifier output is connected to the coax down lead transmission line that comes into the house. Such amplifiers are well sealed and weather protected.

Figure 3-24
Signal amplifiers.

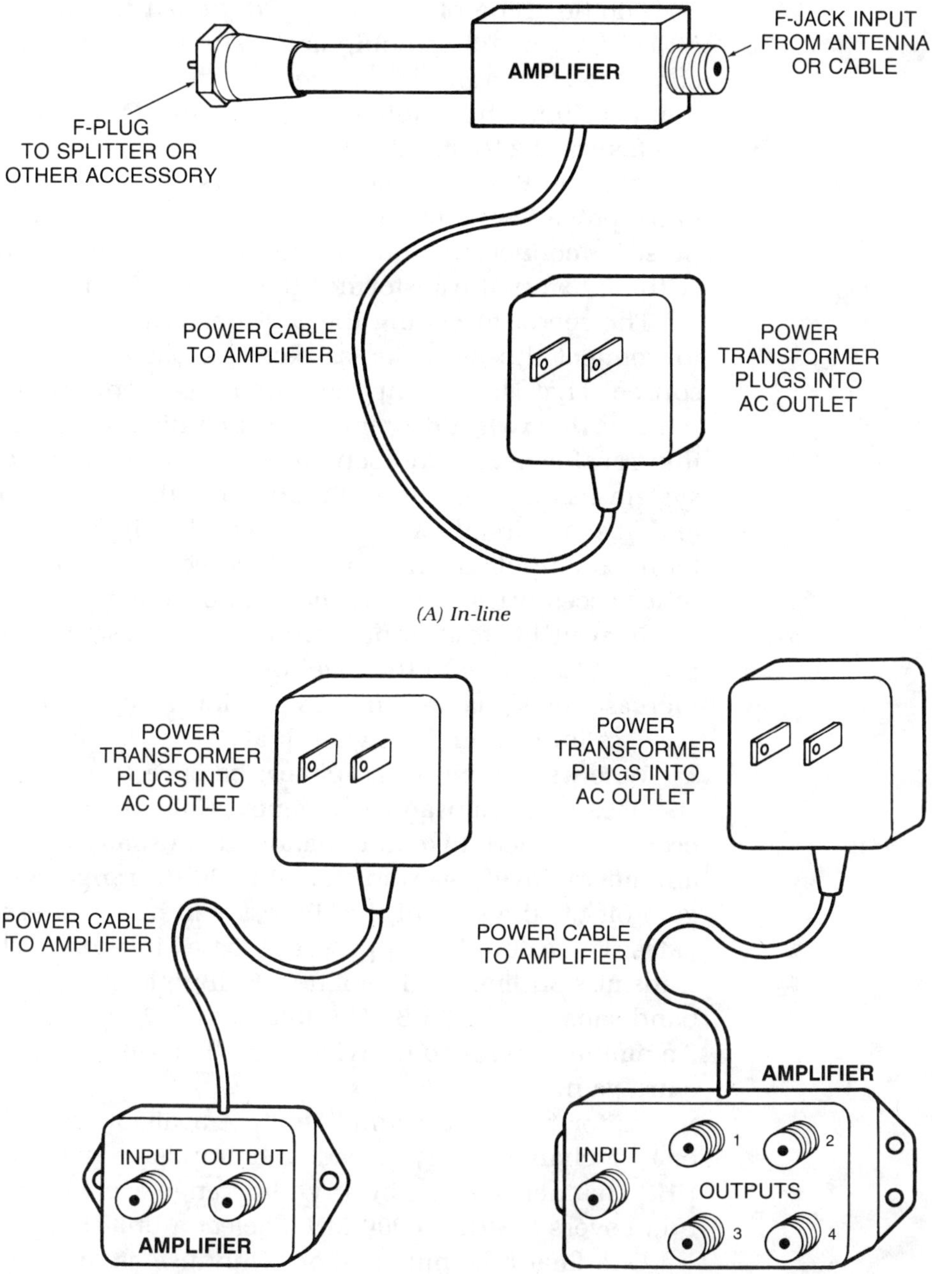

(A) In-line

(B) Single output

(C) Four outputs

Another type of amplifier is designed for mounting on the back of a TV set or other equipment. The input from the TV antenna is connected to the amplifier. The amplifier then has one or more outputs that distribute the amplified signal to the TV set, VCR, or other units. See figures 3-24B and 3-24C.

Like the RF modulator, amplifiers are electronic circuits that require power. All amplifiers have built-in power supply circuitry but must be connected to a source of AC. Typically, they are supplied with an external transformer that is plugged into an AC wall outlet.

The secret to getting the best performance from the amplifier is to connect the amplifier as close as possible to the original signal source. That is, the amplifier should be connected before any components that will add attenuation, including transmission lines, matching transformers, band separators, signal splitters, and the like. In a system using an outdoor TV antenna, the best place to connect the amplifier is directly at the antenna. The signal picked up by the antenna is amplified first, then delivered to the transmission line and other accessories. An amplifier can only work with those signals directly available to its input. Although weak signals are amplified, amplifiers also increase the level of noise and interference. It is best to increase the signal as much as possible so that such outside influence has less impact on the signal quality.

When selecting an amplifier, be sure to look for two important specifications: gain and frequency. Gain is the amount of signal increase provided. Like attenuation, it is usually expressed in dB. Most amplifiers have a gain in the 10 to 30 dB range. Select a unit whose gain offsets the sum of the dB losses in other accessories plus a little extra. For example, suppose the losses in your system are: antenna transmission line, 2 dB; splitter, 3 dB; matching transformer, 2.5 dB; band separator, 1.5 dB. The total loss is $2 + 3 + 2.5 + 1.5 = 9$ dB. An amplifier with 10 dB will offset the 9 dB loss and give you 1 dB of extra gain.

As for frequency, amplifiers are available for VHF channels only or UHF channels only. Broadband amplifiers that cover both VHF and UHF channels and an even wider range are also available. A typical unit covers 5 MHz to 900 MHz. Select a unit that meets your needs.

One final recommendation. Only use as much gain as needed to get a good picture. Excessive gain will overload your set and cause distortion.

Switches and Switchers

A switch is an electronic component used to turn a circuit off or on, or to select desired signals or connections. In video installations, a variety of switches are used not only to help simplify the interconnection of various units, but also to provide a quick and convenient means of changing these interconnections for different functions. Switches allow you to rapidly change the configuration of your system without disconnecting and rewiring cables.

A/B Switches

The most widely used switch is an A/B switch. This switch connects either of two inputs labeled A and B to a single output. Or the switch can route a single input to either of two outputs labeled A or B. Such switches are called signal selectors. A/B switches are available for switching both coax and 300-ohm twin lead.

Figure 3-25 shows the basic electrical connections for such a switch. In electrical and electronic terms, A/B switches are referred to as single pole, double throw (SPDT) switches or double pole, double throw (DPDT) switches.

The SPDT switch has a single arm that can be moved to one of two poles, A or B. The DPDT switch is really two SPDT switches in the same package. The two arms are ganged, or connected, so they operate simultaneously from a single switch handle or lever. SPDT switches are used with coax because only the center conductor needs to be switched. DPDT switches are used to switch twin lead.

Figure 3-26A shows two 75-ohm input signal sources applied to the switch. When the arm of the switch is in the A position, signal source 1 is connected to the output. When the switch arm is in the B position, signal source 2 is applied to the output. Because the shield of the coax is at ground and common to all the inputs and outputs,

Figure 3-25 Basic switch configurations.

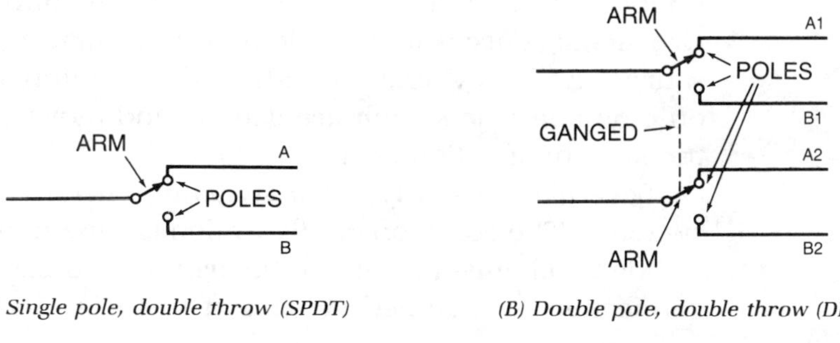

(A) Single pole, double throw (SPDT) *(B) Double pole, double throw (DPDT)*

Figure 3-26
Two ways to use an SPDT A/B switch.

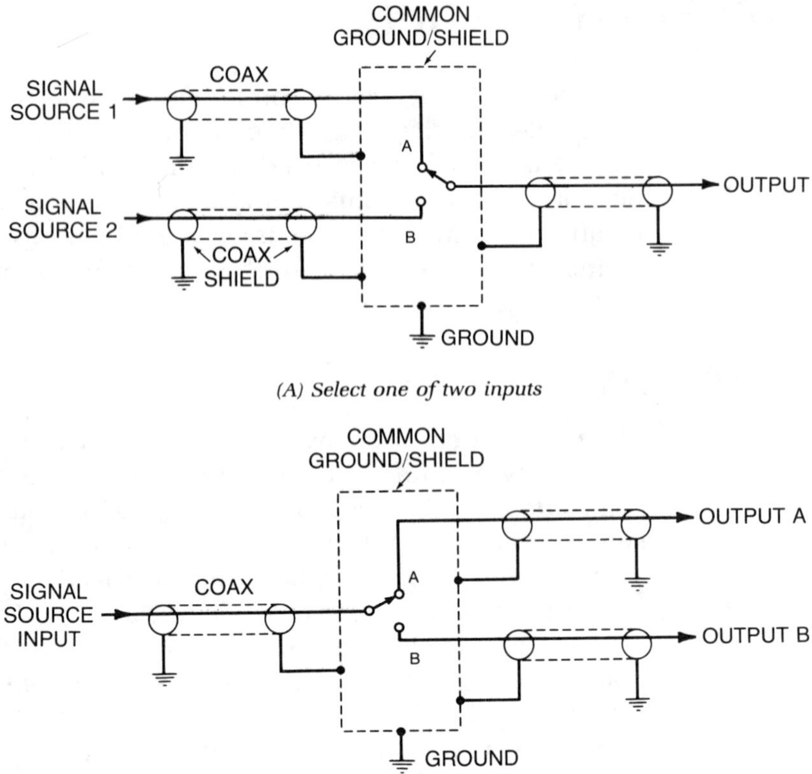

(A) Select one of two inputs

(B) Select one of two outputs

it is not necessary to switch it. Only a single pole, double throw switch is required to switch the coax center conductors.

Figure 3-26B shows the A/B switch used in the opposite direction. Here, a single signal source can be connected to either of two outputs, labeled A or B, depending upon the position of the switch arm.

A/B switches are also available for 300-ohm twin lead. In this case, both of the leads must be switched and, therefore, a double pole, double throw switch is used. Figure 3-27A shows the connections of two signal sources to a single output. In figure 3-27B, a single signal source can be switched to either of two outputs. Note that the poles on a double pole switch are ganged and move simultaneously when the switching action occurs.

Several different types of switching mechanisms are used in A/B switches. The common configuration is shown in figure 3-28A. It uses a slide mechanism to move the arm from the A connection to the B connection. A modified version of this uses a snap-action switching

Figure 3-27 Two ways to switch twin lead with a DPDT A/B switch.

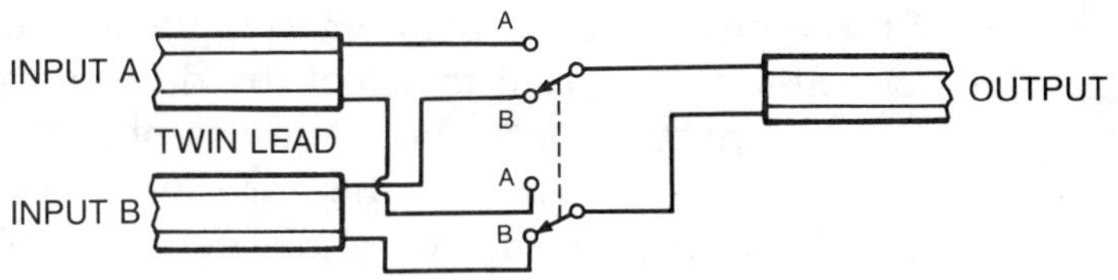

(A) Select one of two inputs

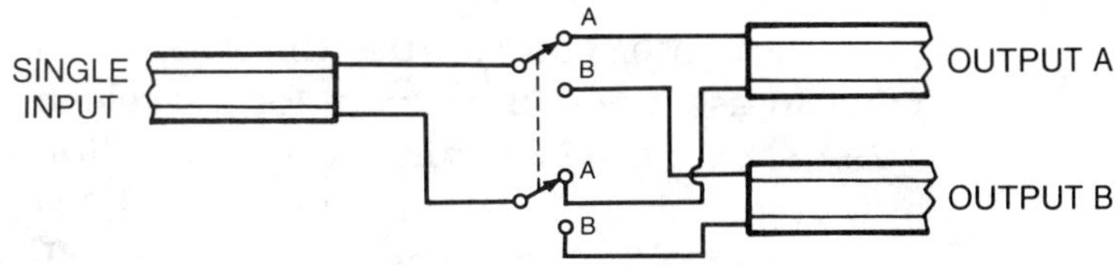

(B) Select one of two outputs

lever like that on a wall light switch. See figure 3-28B. Another type of A/B switch is shown in figure 3-28C. It uses two pushbuttons to select the desired input or output.

A more complex variation of the A/B switch is the A/B/C switch, which contains three inputs and a single output. With this unit, one

Figure 3-28 Types of A/B switches.

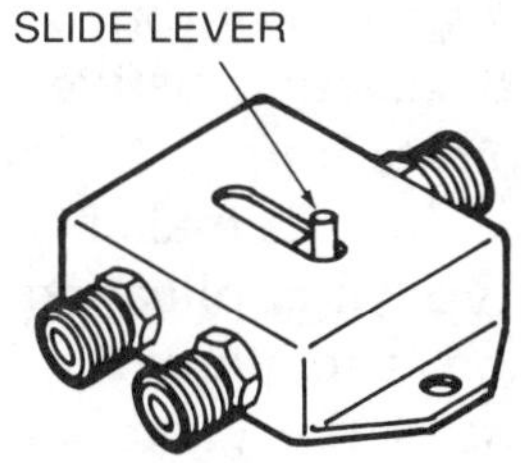

(A) Slide switch

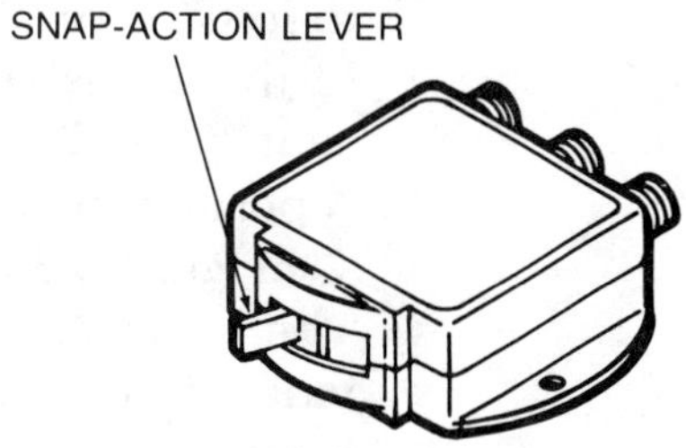

(B) Toggle switch

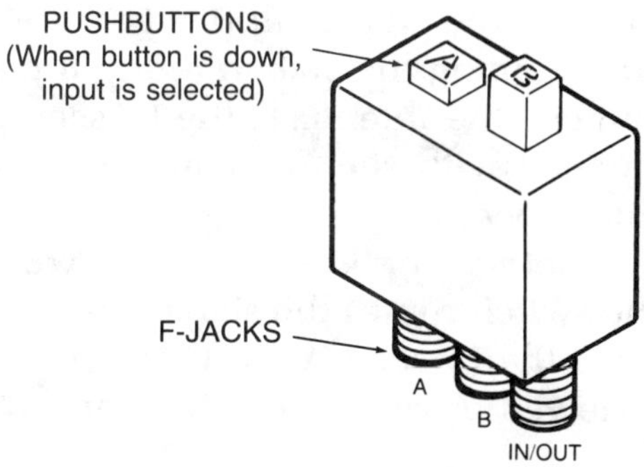

(C) Pushbutton switch

of three inputs can be connected to the single output. Normally, three pushbuttons are used to select the desired input.

An important note about the use of A/B switches is that they are bidirectional. This means they are totally reversible and signals can pass through them in either direction. Thus, inputs can be used as outputs and vice versa. This permits a single switch to be used in many ways depending upon how it is connected.

Sometimes you may want to switch three inputs or three outputs. In such a case, you can use the A/B/C switch just described. If not, you can use two A/B switches for the selection of three signals. Figure 3-29A shows the switching connections for three inputs into a single output. In this application, inputs 1 and 2 are connected to the A and B inputs of switch #1. The output of switch #1 is connected to the A input on switch #2. The third input is connected to the B input on switch #2. The output of switch #2 is signal 1, 2, or 3 depending upon the positions of the switches. To select input 1 or 2, switch #2 must be in the A position. Then signal selection is made with switch #1. To select input 3, switch #2 is placed in the B position.

Figure 3-29B shows how a single input can be connected to any one of three outputs. Actually, the connections are exactly the same as those in figure 3-29A except the figure has been reversed from right to left. To route the input signal to output 1 or 2, switch #1 must be in the B position. Then the selection between 1 and 2 is made with switch #2. To route the input to output 3, simply set switch #1 to the A position. The position of switch #2 makes no difference at this point.

Figure 3-30A shows a typical application for a basic A/B switch. The object in this application is to allow the TV set to be used with a video game. The RF output of the video game is connected to one input of the A/B switch. The other switch input comes from the TV antenna through a matching transformer. The output of the switch goes to the TV set antenna input. When the switch is set to the A position, the signal supplied to the TV set comes directly from the game. Setting the switch to the B position disconnects the game and connects the antenna to the TV set input. Most video games are usually supplied with such a switch, making the purchase of an accessory unit unnecessary.

Another application for an A/B switch is shown in figure 3-30B. The switch routes the signal from a cable TV converter box output to either the TV set or VCR. This switch lets you select whether the signal is viewed or recorded. There are many other similar applications.

Figure 3-29 Using two A/B switches to switch three inputs or outputs.

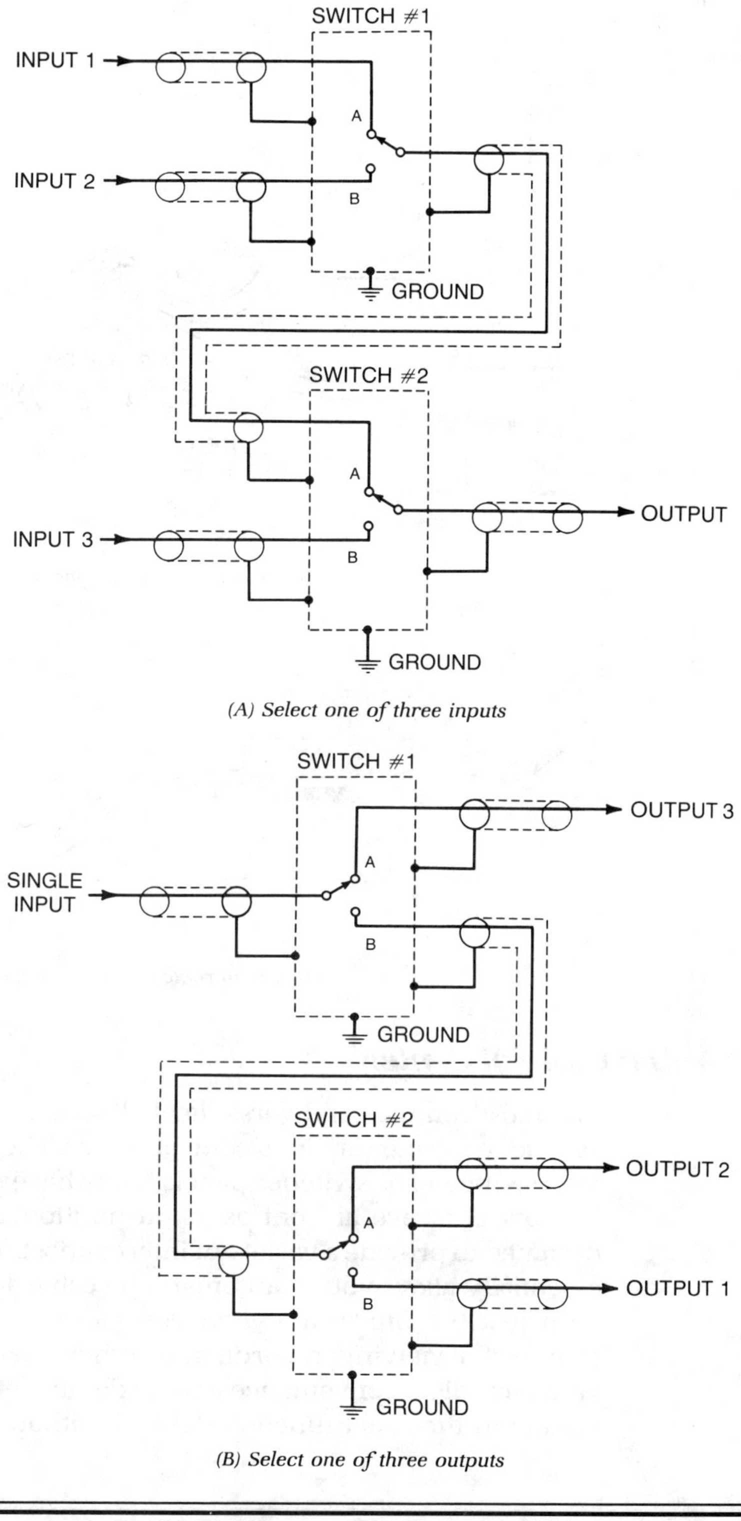

(A) Select one of three inputs

(B) Select one of three outputs

Figure 3-30
A/B switch applications.

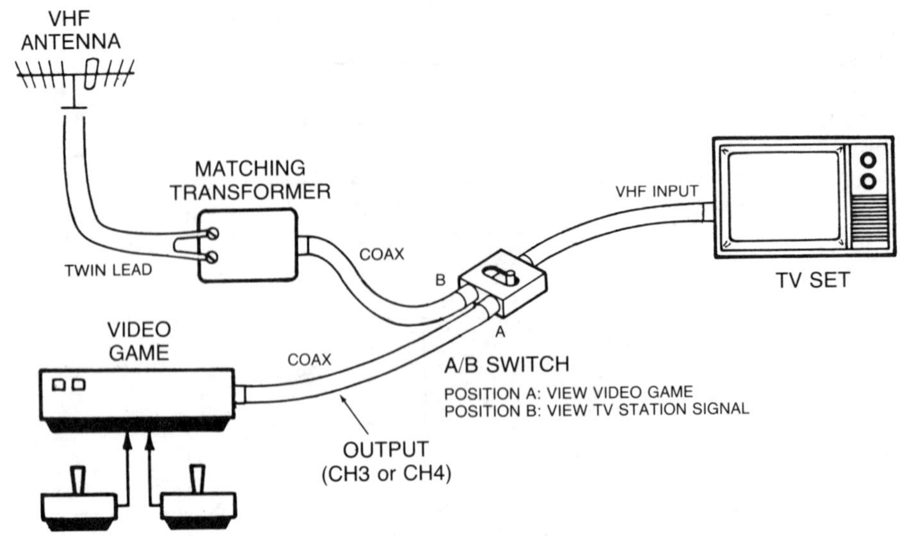

(A) A/B switch used to select one of two inputs

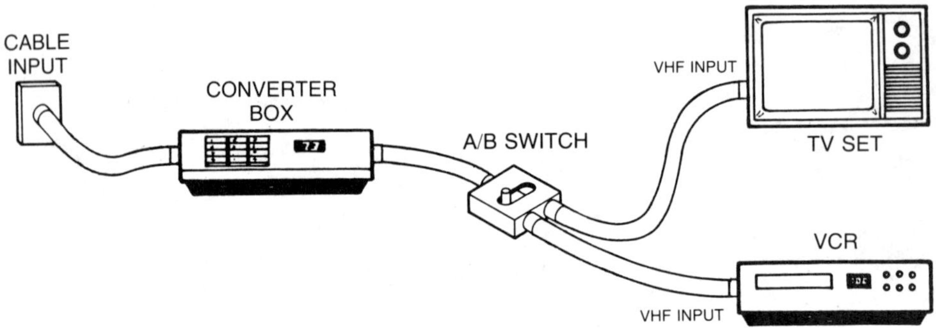

(B) A/B switch used to route one input to one of two outputs

Video Control Center

The most complex and versatile of all switches is a special unit known as a video selector or video control center. These units are also referred to as switchers or switcher panels. They have multiple inputs, multiple outputs, and several front panel pushbutton, lever, or rotary switches designed to provide a variety of interconnections. Such switching units essentially allow you to automate the cable swapping process so that with just the flip of a switch you can direct signals from multiple sources for viewing, recording, or other operations. These units provide incredible convenience and flexibility, letting you configure your system to the exact function desired without hookup problems.

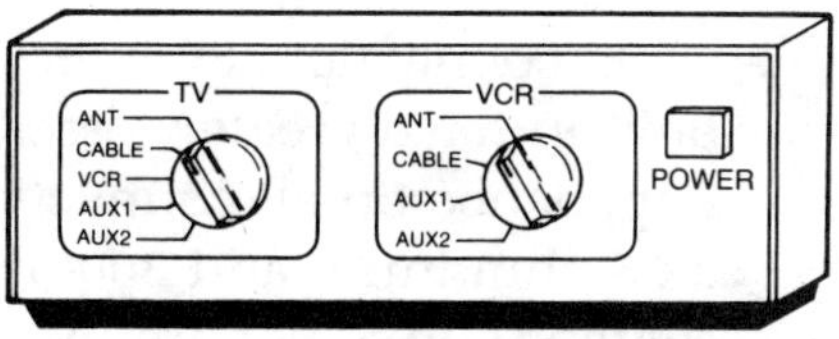

Figure 3-31
Typical video control unit.

Video selector units come in many sizes and configurations. Some feature mechanical pushbutton or rotary front panel switches, and others use more sophisticated electronic switches. A typical unit is shown in figure 3-31. The number of different inputs and outputs also vary considerably. In most cases, these selector units have four or five inputs and two or more different outputs. Front panel switches allow you to essentially route one or more signal sources to one or more outputs. For example, a typical unit accepts antenna, cable, VCR, satellite dish, and video game inputs and uses two rotary switches to route the inputs to the TV set, the VCR, or both.

These selector units work only with coax inputs and outputs, either F-connectors or RCA phono connectors, and maintain full impedance matching. Such switching units introduce attenuation but it is often overcome by an internal amplifier. Some switcher units also contain a built-in RF modulator, thereby eliminating the need for an external unit. For best results, follow the hookup directions in the manual. Because there are so many different variations and configurations, it is not possible to give specific hookup directions here.

Attenuation and Gain

All of the accessories discussed in this chapter, except the amplifier, introduce signal attenuation. They perform their specific functions but at the cost of loss in signal level. Some units introduce minimum attenuation, which is not noticed. Others insert significant loss, which shows up as a snowy picture. Using multiple accessories can also cause enough loss to reduce picture quality.

Some system accessories affect signal strength more than others. The following list orders accessories from least loss to greatest loss.

1. Connectors and adapter jacks. Connectors and adapter jacks cause some slight signal loss. They do, however, allow easy changes in any hookup.
2. Signal switches. Signal switches do not split or separate the signal. But they do not pass a signal as effectively as a plain

cable connection. Switches that have matching transformers built in have greater losses than switches that are the same impedance on all connections. In other words, a switch with a 75-ohm input and 300-ohm output has a greater loss than a switch with a 75-ohm input and a 75-ohm output.

3. Transmission line. Both coax and twin lead introduce some loss. Coax attenuation is significantly more than twin lead attenuation. The attenuation is directly proportional to length. Short cables have negligible loss. But long cables such as antenna lead-ins of 20 feet or more have a major loss. As a general rule, keep all cable lengths as short as practical.
4. Band separator. Band separators separate the UHF, VHF, and FM bands. Each band has only one output, so there is no signal splitting, just separation. When the selected band is coupled with a designated output, separation with minimum signal loss of any band results. The band splitter minimizes interference and tuning problems even though some loss occurs in the process of separating the bands.
5. Filters and traps. Filters eliminate undesirable interference but cause insertion loss during this process.
6. Signal splitters. Signal splitters divide an input signal into more than one output. At the same time, they shield the system from interference and match the impedances of the inputs and outputs. Splitters introduce major attenuation, so use them sparingly.

Gain and loss are expressed in decibels, dB. A decibel is one tenth of a bel, which is a unit commonly used to indicate increases or decreases in sound or electronic signal level.

Decibels are derived from the ratio of the signal input and output power. If the output is less than the input, a loss is present and the dB value is negative. If the output is greater than the input, there is gain or amplification and the dB value is positive. Table 3-1 shows the dB values associated with common input/output power ratios. For example, if the output is half the input, the input must be two times the output. This gives a power gain ratio of 2, which corresponds to 3 dB. In this example, it is a loss so we express it as -3 dB.

If the output is ten times the input, you can see by the table that this represents 10 dB of gain. The curve in figure 3-32 is a plot of the data in table 3-1.

Table 3-1.
Decibels Versus Power Ratio

Decibels (dB)	Power Ratio (Gain)	
1	1.25	(.25 more signal power)
2	1.5	(.50 more signal power)
3	2	(2 times original signal power)
4	2.5	(2.5 times original signal power)
5	3	(3 times original signal power)
6	4	
7	5	
8	6.33	
9	8	
10	10	
11	12.5	
12	16	
13	20	
14	25	
15	31.5	
16	40	
17	50	
18	63	
19	79	
20	100	

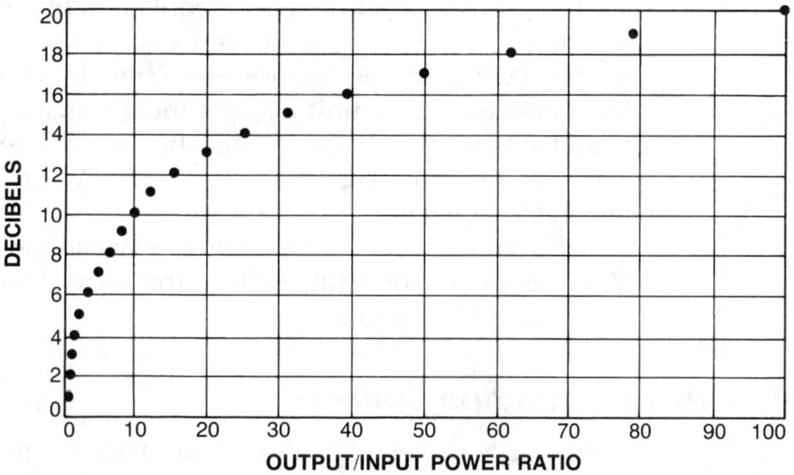

Figure 3-32
Graph of table 3-1.

To find the total loss or gain in a system, just add all the losses and gains. If accessories introduce losses of 1.8, 2.3, 3.5, and .7 dB, the total loss is −8.3 dB.

Total gain is computed the same way, by adding the dBs of amplification. Suppose an antenna has a 10 dB gain and an amplifier adds 14 dB. The total gain is 10 + 14 = 24 dB.

The total gain or loss is found by adding the gains and losses. Using the gain and loss figures from the previous examples, we find a total of $+24 - 8.3 = 15.7$ dB gain. If the gain was $+12$ dB and the total attenuation -18.5 dB, the total would be $+12 - 18.5 = -6.5$ dB or a 6.5 dB loss.

Special Accessories

The previous sections covered most of the common accessories you will use in video hookups. But, as you will discover, many others perform unique functions that may or may not be important to you. We do not cover the hookup of these special units. Use the instructions supplied with the unit.

Closed Caption Box

A closed caption box is a special converter unit. Many TV programs are closed captioned. That is, they are accompanied by subtitles printed at the bottom of the screen. Those with hearing impairment can read the dialog.

Closed captioned programs contain subtitles in special coded signals transmitted along with the usual picture and sound signals. Ordinary TV sets simply ignore these signals. To use them, a converter box is needed. This unit accepts the TV signal like a cable converter box, decodes the subtitle signals, then adds these to the picture signal. The composite TV signal is sent to the TV set (or VCR). The subtitles are then displayed.

The converter box is connected between the antenna or cable input and the TV set or VCR. Follow the special installation instructions supplied with the unit.

Video Distribution System

Video distribution units allow you to send signals to two or more TV sets in other rooms. A typical unit handles four TV sets. The distribution box contains switches and amplifiers that allow any set to use any available signal source. All sets can share a source or each set can use a different source. One set can view a signal from the antenna, another set can view a premium cable channel, while another plays a prerecorded tape.

The distribution unit is usually installed at the main TV set, where all sources are available. The signals to the other sets are transmitted by coax cables.

Video Processor

A video processor is an electronic unit that operates on (processes) video signals to improve their quality. Video processors, also known as image enhancers, are used with a VCR.

Recording or copying video tapes reduces their quality. The color may no longer be true, the contrast may be poor, and noise may be introduced. A video processor corrects all these problems. When playing, recording, or copying a tape, you can pass the video signal through the processor and enhance the signal by adjusting various front panel controls. The result is a signal of far greater quality than the original.

The video processor is connected between two VCRs when copying a tape or between the VCR and the TV set when playing a tape. When recording, the image enhancer is placed between the signal source and the VCR. Obviously, special hookups are required. Most video processors have internal switches and accessories that help simplify installation. Be sure to follow the manufacturer's instructions.

AC Power Connections

An often overlooked part of a video system is the AC power wiring. You take it for granted that all your equipment must be plugged into an AC outlet. This is no problem if all you have is a TV set and VCR. Most AC outlets have two receptacles to accept plugs. But what happens if you have three or more pieces of equipment? Naturally, you need an extension cord with multiple receptacles or a multireceptacle block that plugs into the AC output and accepts two or more plugs. Even with such accessories, the hookup can be messy and even dangerous.

The best solution to this problem is to purchase an AC power strip usually sold for personal computer systems. A typical unit is shown in figure 3-33. The heavy-duty power cord plugs into the AC outlet. Then all your video accessories plug into the AC outlets on the strip. Units are available with 4 to 12 receptacles. Most units also contain an on/off switch and a pilot light. Some strips also have a fuse or circuit breaker for protection.

Many strips used on computer systems include protective components or circuits. Most feature spike suppression and overload prevention. This eliminates the high-voltage pulses or high-voltage surges that often occur on power lines. They can damage computer equip-

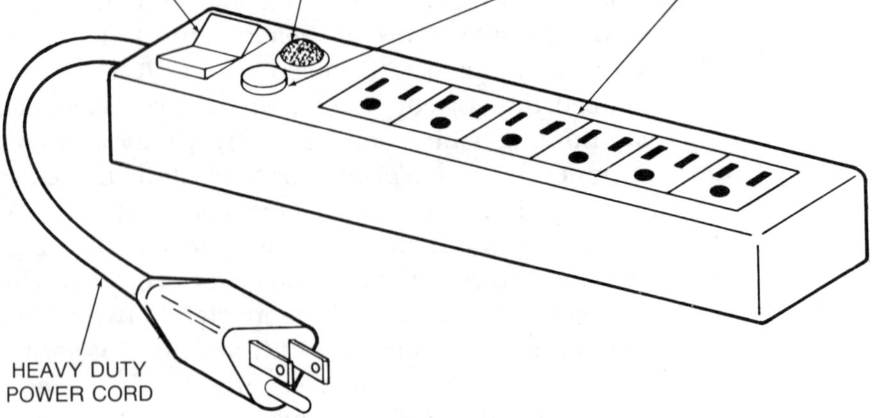

Figure 3-33 AC power strip.

ment or any other electronic units, including TV sets and VCRs. It is strongly recommended that the strip you use contain such spike and overload prevention devices. In this way, your investment in video equipment is protected against AC power line troubles. Most units also contain power line filters that help remove noise and other spurious signals picked up by power lines. This provides the added benefit of less noise and interference and better picture quality.

4 Hookup Preparation

Now that you know all about your equipment and interconnection accessories, you are ready to make a hookup. You need to select a hookup before collecting the parts and performing the wiring.

Assumptions

The following assumptions are used in all the hookups:

1. At least a TV set and a VCR are used in each system.
2. No computers are used in the hookups. If a computer is to be used productively, it should be located on its own desk, away from entertainment systems. TV screens are unsuitable for most computer displays.
3. If a satellite TV receiver is used, the cable TV input is not used.
4. If cable TV is available, a satellite TV receiver is not used. (It is not required because most signals are available on cable.)

Selecting the Hookup

The hookup you select will be governed by the equipment you have and what functions you want to perform. You will have at least a TV set and a VCR. Further, you will want to be able to record one signal source while viewing another. Not all hookups permit this, but that is a goal. In general, you want the greatest flexibility, most advantages, and least disadvantages. Advantages and disadvantages are listed when necessary to help you decide.

Before selecting a hookup, read this chapter for general guidelines. Then refer to Table 4-1 to choose a hookup. This quick reference guide will point you to those hookups appropriate to your equipment. Also refer to the contents of chapters 5, 6, and 7 for an overview of the hookups available.

Read each hookup appropriate to your situation. Pay attention to two things:

1. Is your TV set, VCR, or both cable-ready or not? Check your manuals if you are not sure.
2. Is your primary signal source a TV antenna or a cable system? These factors will narrow your choices.

Next, read the advantages and disadvantages, if listed. Then select the hookup appropriate to your situation.

Each hookup specifies the equipment and accessories needed. Exact step-by-step hookup procedures are listed. These supplement the clear diagram of each hookup. Each hookup concludes with step-by-step operational details for each function the system can perform.

Table 4-1. Guide to Hookups in Chapters 5, 6, and 7

Equipment	Hookups
TV set and VCR only	
Non-cable-ready	1, 2, 3, 4, 5
Cable-ready	9
TV set and VCR plus one or more of the following	
Satellite receiver	6, 7, 8
Cable converter-decoder box	
Cable-ready TV, cable-ready VCR	11, 12
Cable-ready TV, non-cable-ready VCR	13
Non-cable-ready TV, cable-ready VCR	14, 15, 16
Non-cable-ready TV, non-cable-ready VCR	10, 17
Videodisc player	8
Second VCR	19, 22
Second cable converter-decoder	18
Stereo decoder	21
Video camera	23
Stereo hi-fi system	20, 21
Video game	24
Two-cable signal source	15, 19

Preparing the Signal Source

Your primary signal source will be either a TV antenna or cable. Your first step is to prepare the source. You will use one or the other, but not both.

The antenna can be outside, in the attic, or just a set of rabbit ears on top of the TV set. It can be VHF, UHF, or both. Or it can be a master antenna system in your apartment building or condo. Whatever the configuration, it will use either 300-ohm twin lead or 75-ohm coax. The antenna's transmission line can be terminated at a convenient wall outlet or can be just a cable through a hole in the wall.

VHF (Channels 2 Through 13)

VHF antennas use either 300-ohm twin lead or 75-ohm coax. Be sure to identify the type of VHF connection to your TV so that the impedance of the leads matches the impedance marked on the jacks. If the VHF connection at the TV is a threaded sleeve with a hole in the middle, it is a 75-ohm F-jack. If the VHF connection at the TV is a pair of screws, it is a 300-ohm twin lead connection. You will need to use matching transformers if the antenna leads and the TV set or VCR inputs are different impedances.

UHF (Channels 14 Through 83)

The 300-ohm twin lead is usually used for UHF antenna transmission lines. No matching transformers are needed to connect 300-ohm twin lead to 300-ohm UHF inputs on your TV set or VCR. Two screw terminals are the standard connectors for the twin lead, at the VCR and at the back of the TV. Some newer TV sets have a single 75-ohm coax input for a composite VHF/UHF signal from the antenna or cable system. If there are two sets of two screws on the back of the TV, one pair is for the UHF and the other is for the VHF input.

Note: Use the correctly paired screws for connections from the VCR to the TV set. UHF must be connected to UHF, and VHF must be connected to VHF.

Combo UHF/VHF/FM

If UHF and VHF signals come from an antenna on one cable, a band separator is needed before the cable is connected to the TV or VCR antenna inputs. Select a band separator with the appropriate input and output impedances and connections to match your antenna and equipment.

If your source is a cable from your local cable company, it will be 75-ohm coax. One cable is common, but the newer, larger systems have two cables, one for standard channels and the other for scrambled premium channels. The cable or cables must have F-plugs. Usually, the cable goes to a cable converter-decoder box. But in systems with a cable-ready TV set or VCR, the cable can bypass the cable box. The cable signal can also pass through splitters or A/B switches prior to being routed to the equipment.

Preparing TV and VCR Inputs

To simplify the hookups in the following chapters, prepare the TV, VCR, or both for 75-ohm connection at the VHF input, if it is not already equipped with a 75-ohm F-jack. Taking the time to make this conversion now will save time later. Use matching transformers to match between different impedances on the equipment, cables, and accessories as necessary.

If your TV has only screw terminal twin lead connectors for twin lead connection to the VHF input, connect a 75-ohm-to-300-ohm matching transformer to the TV VHF input, as illustrated in figure 4-1. Use one on the VCR antenna input if an F-jack is not used.

To connect a 75-ohm-to-300-ohm matching transformer to the TV or VCR 300-ohm VHF screw terminals, loosen the screws and remove any 300-ohm twin lead. Connect the 300-ohm twin lead of the matching transformer. Place one lead under each screw and tighten both screws securely.

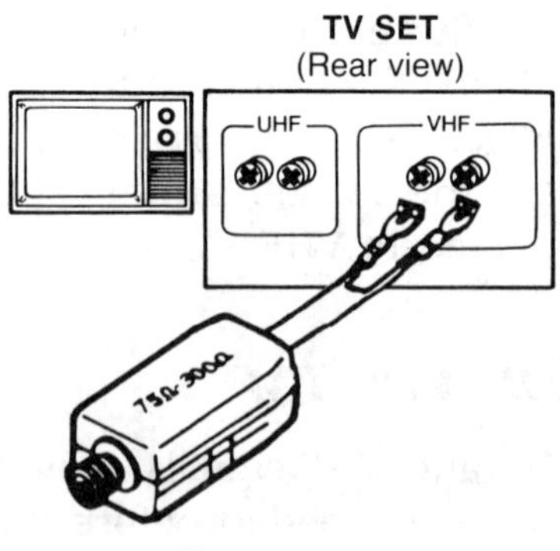

Figure 4-1 Matching transformer connection.

Note: The 75-ohm F-connector on the end of the transformer opposite the 300-ohm twin lead is now ready to be the 75-ohm TV VHF input. Other matching transformers can also make other hookups faster and easier. They can also adapt rabbit ear antennas or other twin lead cable to the 75-ohm VHF ANT jack of the VCR, as shown in figure 4-2.

Figure 4-3 shows how to use a band separator to convert the input from a VCR coax output into the VHF (coax) and UHF (twin lead) connections required by the TV set.

Hookup Rules

As you read the hookups in the next chapters, follow these basic rules:

1. For video, audio, and RF signals, connect outputs to inputs and vice versa. For example, connect the video output of a VCR to the video input on a TV set or monitor. Do *not* connect output to output or input to input.
2. Match impedances. Use impedance matching transformers or baluns for 75-ohm-to-300-ohm transformation as described in the previous section.
3. Use only one input signal source. For example, if cable TV is available, eliminate the antenna and associated cables. Over-the-air signals are usually available on the cable anyway. This simplifies connections.
4. Minimize the use of accessories such as band separators and splitters to reduce signal attenuation. Choose the simplest hookup that does what you want.
5. Keep cable lengths short to minimize cable tangle, clutter, and signal attenuation.

Figure 4-2
Connection for twin lead and 75-ohm jack.

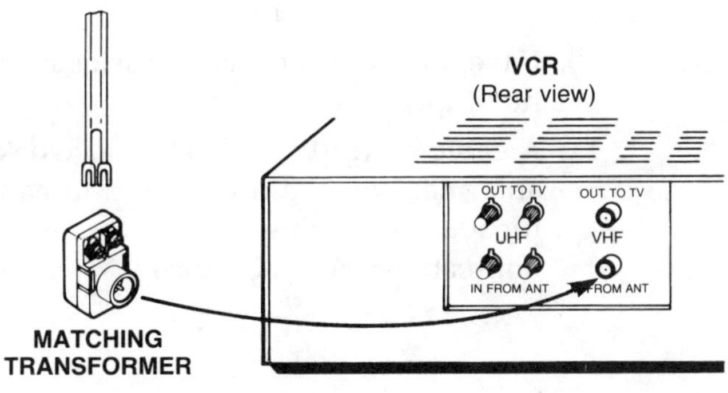

Figure 4-3
Hookup to convert TV to 75-ohm UHF-VHF connection.

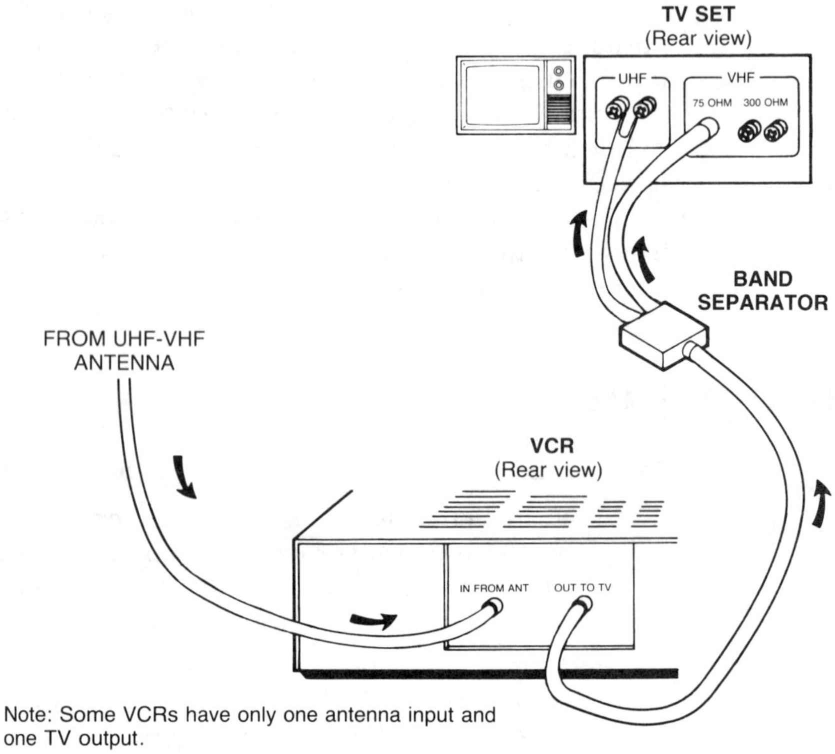

6. Coil unused cable and secure with a twist-tie for neatness.
7. Support cables where possible and do not pull or stretch them. This minimizes stress on the connectors.
8. When hooking up speakers, connect + terminals to + terminals and − terminals to − terminals. This ensures proper phasing. Use color-coded cable or other cable markings to identify each wire in the cable.
9. For safety, turn off power or unplug all equipment before making the hookup.
10. Have all cables and accessories ready before making the hookup.
11. A cable-ready VCR or TV will work with either cable or an antenna. You are not limited to cable use. In hookups 1 through 8, a cable-ready VCR or TV can be substituted for a non-cable-ready VCR or TV.

5 Simple Hookups

If your system is relatively simple, this chapter probably contains an appropriate hookup. The hookups described here involve a TV set, a VCR, plus either a cable box or satellite receiver. These basic hookups do not offer full versatility, but you may not want or need it. For maximum flexibility and full performance capability, use one of the more complex hookups in chapter 6.

Hookup 1—VCR to TV

Equipment
Non-cable-ready VCR, non-cable-ready TV, no antenna or cable.

Accessories
One 75-ohm coaxial cable with F-plugs. Place equipment in position and note cable length needed before purchasing cable.

Hookup Procedure
Refer to figure 5-1 before hooking up equipment. If any equipment has already been installed, check each connection before making any changes. Some connections may not need to be changed.

1. Connect one end of a 75-ohm cable to the VHF OUT on the VCR.

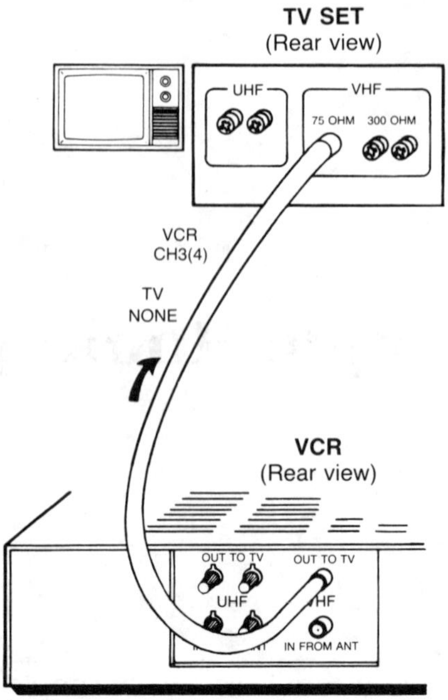

Figure 5-1
Hookup 1.

2. Connect the other end of the cable to the 75-ohm VHF antenna input on the TV. *Note:* If you do not find a 75-ohm VHF antenna input on your TV, prepare the TV for a 75-ohm connection at the VHF input. See chapter 4.

To View a Prerecorded Tape

1. Be sure both power switches are on.
2. Set the TV/VCR switch on the VCR to VCR.
3. Tune the TV to the output channel of the VCR (channel 3 or 4).
4. Press the VCR play button.

Hookup 2—VHF Antenna to VCR

Equipment

Non-cable-ready VCR, non-cable-ready TV, VHF antenna.

Accessories

One 75-ohm coaxial cable. Place equipment in position and note cable length needed before purchasing cable.

Hookup Procedure

Refer to figure 5-2 before hooking up equipment. If any equipment has already been installed, check each connection before making any changes. Some connections may not need to be changed.

1. Connect one end of a 75-ohm cable to VHF OUT TO TV on the VCR. Connect the other end of the cable to the 75-ohm VHF antenna input on the TV. *Note:* If you do not find a 75-ohm VHF antenna input on your TV, prepare the TV for a 75-ohm connection at the VHF input. See chapter 4.
2. Connect the 75-ohm VHF antenna coax to the 75-ohm VHF IN jack on the VCR.

To View a Prerecorded Tape

1. Be sure both power switches are on.
2. Set the TV/VCR switch on the VCR to VCR.
3. Tune the TV to the output channel of the VCR (channel 3 or 4).
4. Press the play button on the VCR.

Figure 5-2
Hookup 2.

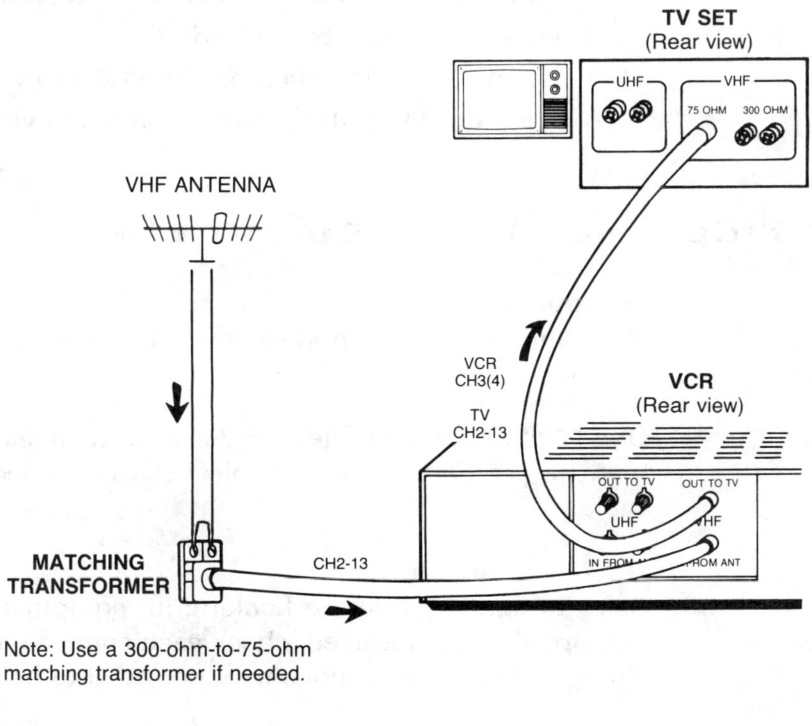

To View the TV Without Recording

1. Turn the TV power on. Turn the VCR power off.
2. Select the channel to view with the TV tuner.

To Record and View the Same Channel

1. Be sure all power switches are on.
2. Tune the VCR to the channel to record.
3. Set the TV/VCR switch on the VCR to VCR.
4. Tune the TV to the output channel of the VCR (channel 3 or 4).
5. View the TV to verify the channel to record.
6. Press the record button(s) on the VCR.

To View One Channel While Recording Another Channel

1. Be sure all power switches are on.
2. Tune the VCR to the channel to record.
3. Set the TV/VCR switch on the VCR to VCR.
4. Tune the TV to the output channel of the VCR.
5. View the TV to verify the channel to record.
6. Press the record button(s) on the VCR.
7. Set the TV/VCR switch on the VCR to TV.
8. Tune the TV to the desired channel to view.

Hookup 3—UHF Antenna to VCR

Equipment

Non-cable-ready VCR, non-cable-ready TV, UHF antenna.

Accessories

One 75-ohm coaxial cable, one 300-ohm twin lead cable. Place equipment in position and note cable lengths needed before purchasing cable.

Hookup Procedure

Refer to figure 5-3 before hooking up equipment. If any equipment has already been installed, check each connection before making any changes. Some connections may not need to be changed.

Figure 5-3
Hookup 3.

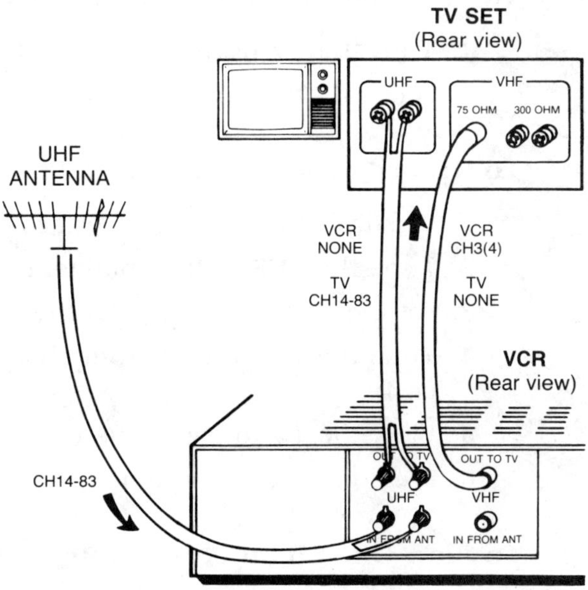

Note: To play tapes, VHF OUT must be connected to the TV VHF antenna input.

1. Connect one end of a 75-ohm cable to the VHF OUT jack on the VCR. Connect the other end of the cable to the 75-ohm VHF antenna input on your TV. *Note:* If you do not find a 75-ohm VHF antenna input on your TV, prepare the TV for a 75-ohm connection at the VHF input. See chapter 4. The VHF output is needed for most VCR output channels of channel 3 or 4.
2. Connect one end of the 300-ohm twin lead to the UHF OUT terminals on the VCR. Connect the other end of the twin lead to the 300-ohm UHF antenna input terminals on the TV.
3. Connect the 300-ohm twin lead from the UHF antenna to the UHF IN terminals on the VCR.

To View a Prerecorded Tape

1. Be sure both power switches are on.
2. Set the TV/VCR switch on the VCR to VCR.
3. Tune the TV to the output channel of the VCR (channel 3 or 4).
4. Press the play button on the VCR.

To View the TV Without Recording

1. Turn the TV power on. Turn the VCR power off.
2. Select the channel to view with the TV tuner.

To Record and View the Same Channel

1. Be sure all power switches are on.
2. Tune the VCR to the channel to record.
3. Set the TV/VCR switch on the VCR to VCR.
4. Tune the TV to the output channel of the VCR (channel 3 or 4).
5. View the TV to verify the channel to record.
6. Press the record button(s) on the VCR.

To View One Channel While Recording Another Channel

1. Be sure all power switches are on.
2. Tune the VCR to the channel to record.
3. Set the TV/VCR switch on the VCR to VCR.
4. Tune the TV to the output channel of the VCR (channel 3 or 4).
5. View the TV to verify the channel to record.
6. Press the record button(s) on the VCR.
7. Set the TV/VCR switch on the VCR to TV.
8. Tune the TV to the desired channel to view.

Hookup 4—VHF and UHF Antennas to VCR

Equipment
Non-cable-ready VCR, non-cable-ready TV, VHF antenna, UHF antenna.

Accessories
One 75-ohm coaxial cable, one 300-ohm twin lead cable. Place equipment in position and note cable lengths needed before purchasing cable.

Hookup Procedure

Refer to figure 5-4 before hooking up equipment. If any equipment has already been installed, check each connection before making any changes. Some connections may not need to be changed.

1. Connect one end of a 75-ohm cable to the VHF OUT jack on the VCR. Connect the other end of the cable to the 75-ohm VHF antenna input on the TV. *Note:* If you do not find a 75-ohm VHF antenna input on your TV, prepare the TV for a 75-ohm connection at the VHF input. See chapter 4.
2. Connect one end of the 300-ohm twin lead to the UHF OUT terminals on the VCR. Connect the other end of the twin lead to the 300-ohm UHF antenna input terminals on the TV.
3. Connect the 300-ohm twin lead from the UHF antenna to the UHF IN terminals on the VCR.
4. Connect the 75-ohm cable from the VHF antenna to the VHF IN jack on the VCR.

Figure 5-4
Hookup 4.

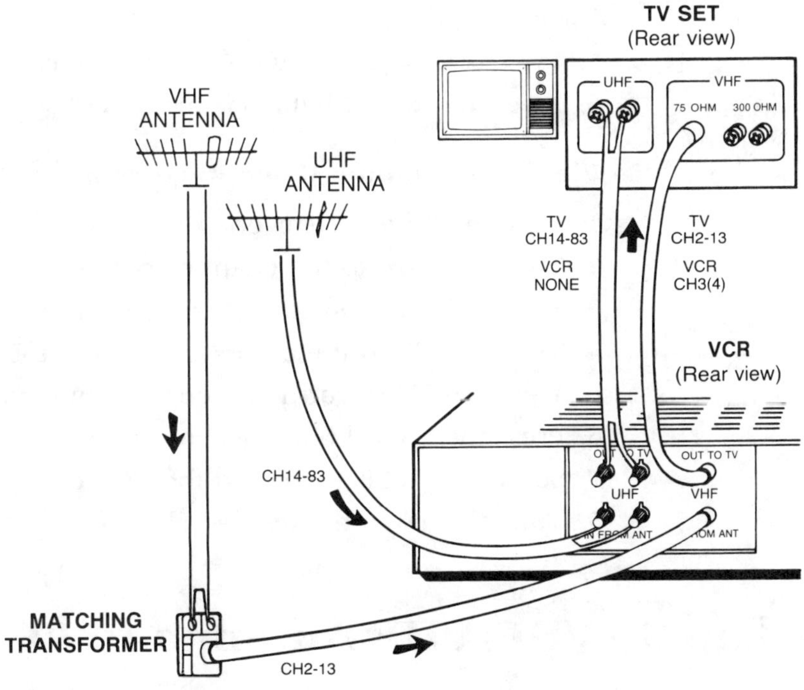

Note: Use a 300-ohm-to-75-ohm matching transformer if needed.

To View a Prerecorded Tape

1. Be sure both power switches are on.
2. Set the TV/VCR switch on the VCR to VCR.
3. Tune the TV to the output channel of the VCR (channel 3 or 4).
4. Press the play button on the VCR.

To View the TV Without Recording

1. Turn the TV power on. Turn the VCR power off.
2. Select the channel to view with the TV tuner.

To Record and View the Same Channel

1. Be sure all power switches are on.
2. Tune the VCR to the channel to record.
3. Set the TV/VCR switch on the VCR to VCR.
4. Tune the TV to the output channel of the VCR (channel 3 or 4).
5. View the TV to verify the channel to record.
6. Press the record button(s) on the VCR.

To View One Channel While Recording Another Channel

1. Be sure all power switches are on.
2. Tune the VCR to the channel to record.
3. Set the TV/VCR switch on the VCR to VCR.
4. Tune the TV to the output channel of the VCR.
5. View the TV to verify the channel to record.
6. Press the record button(s) on the VCR.
7. Set the TV/VCR switch on the VCR to TV.
8. Tune the TV to the desired channel to view.

Hookup 5—VHF/UHF Antenna to VCR

Equipment
Non-cable-ready VCR, non-cable-ready TV, combination UHF and VHF antenna.

Accessories

One 75-ohm coaxial cable, one 300-ohm twin lead cable, one UHF-VHF band separator with correct input and output impedances to match from the antenna to the ANT input connections on the VCR. Place equipment in position and note cable lengths needed before purchasing cable.

Hookup Procedure

Refer to figure 5-5 before hooking up equipment. If any equipment has already been installed, check each connection before making any changes. Some connections may not need to be changed.

1. Connect one end of the 75-ohm cable to the VHF OUT jack on the VCR. Connect the other end to the 75-ohm VHF antenna input jack on the TV. *Note:* If you do not find a 75-ohm VHF antenna input on your TV, prepare the TV for a 75-ohm connection at the VHF input. See chapter 4. The VHF output is needed for most VCR output channels of channel 3 or 4.

Figure 5-5
Hookup 5.

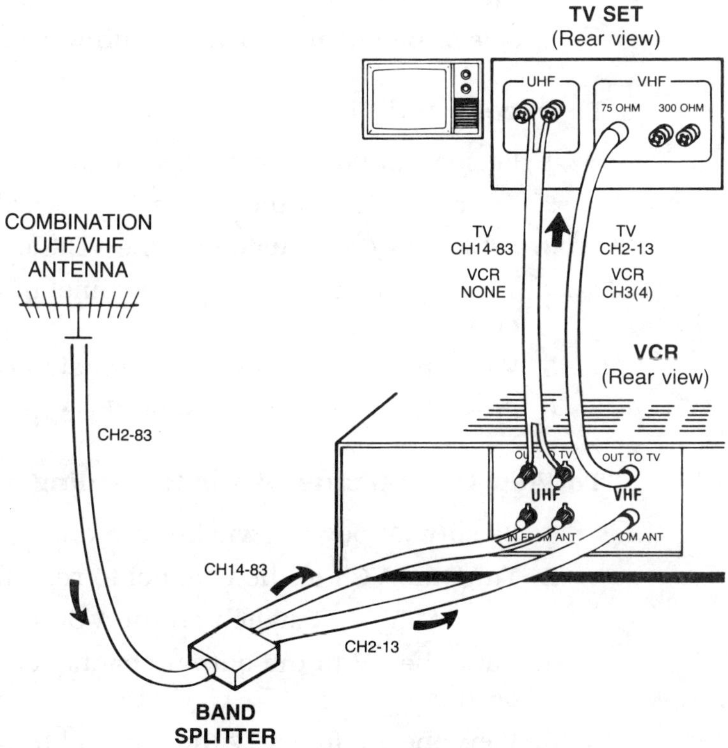

2. Connect one end of the 300-ohm twin lead to the UHF OUT terminals on the VCR. Connect the other end of the twin lead to the 300-ohm UHF antenna input terminals on the TV.
3. Connect the 300-ohm twin lead from the UHF output of the band separator to the UHF IN jack on the VCR.
4. Connect the 75-ohm cable of the VHF output of the band separator to the VHF IN jack on the VCR.
5. Connect the cable from the antenna to the input of the band separator.

To View a Prerecorded Tape

1. Be sure both power switches are on.
2. Set the TV/VCR switch on the VCR to VCR.
3. Tune the TV to the output channel of the VCR.
4. Press the play button on the VCR.

To View the TV Without Recording

1. Turn the TV power on. Turn the VCR power off.
2. Select the channel to view with the TV tuner.

To Record and View the Same Channel

1. Be sure all power switches are on.
2. Tune the VCR to the channel to record.
3. Set the TV/VCR switch on the VCR to VCR.
4. Tune the TV to the output channel of the VCR (channel 3 or 4).
5. View the TV to verify the channel to record.
6. Press the record button(s) on the VCR.

To View One Channel While Recording Another Channel

1. Be sure all power switches are on.
2. Tune the VCR to the channel to record.
3. Set the TV/VCR switch on the VCR to VCR.
4. Tune the TV to the output channel of the VCR (channel 3 or 4).
5. View the TV to verify the channel to record.

6. Press the record button(s) on the VCR.
7. Set the TV/VCR switch on the VCR to TV.
8. Tune the TV to the desired channel to view.

Hookup 6—Satellite Receiver with Block Converter to VCR

Equipment
Non-cable-ready VCR, non-cable-ready TV, satellite receiver, satellite-receiver block converter with UHF channels output on either a 300-ohm twin lead or a 75-ohm coax.

Accessories
One 75-ohm coaxial cable, one 300-ohm twin lead cable, 75-ohm-to-300-ohm matching transformer. Place equipment in position and note cable lengths needed before purchasing cable.

Advantages
Can record and view different channels from the same satellite. Can use the programmable timer feature of the VCR.

Disadvantage
Cannot record a channel from one satellite while viewing a channel from another satellite.

Hookup Procedure
Refer to figure 5-6 before hooking up equipment. If any equipment has already been installed, check each connection before making any changes. Some connections may not need to be changed.

1. Connect one end of a 75-ohm cable to the VHF OUT jack on the VCR. Connect the other end of the coax to the 75-ohm VHF antenna input jack on the TV. *Note:* If you do not find a 75-ohm VHF antenna input on your TV, prepare the TV for a 75-ohm connection at the VHF input. See chapter 4.
2. Connect one end of a 300-ohm twin lead to the UHF OUT terminals on the VCR. Connect the other end of the twin lead to the 300-ohm UHF antenna input terminals on the TV.
3. Connect the block converter output jack to the 75-ohm input of a matching transformer with a coax cable with F-plugs.

Figure 5-6
Hookup 6.

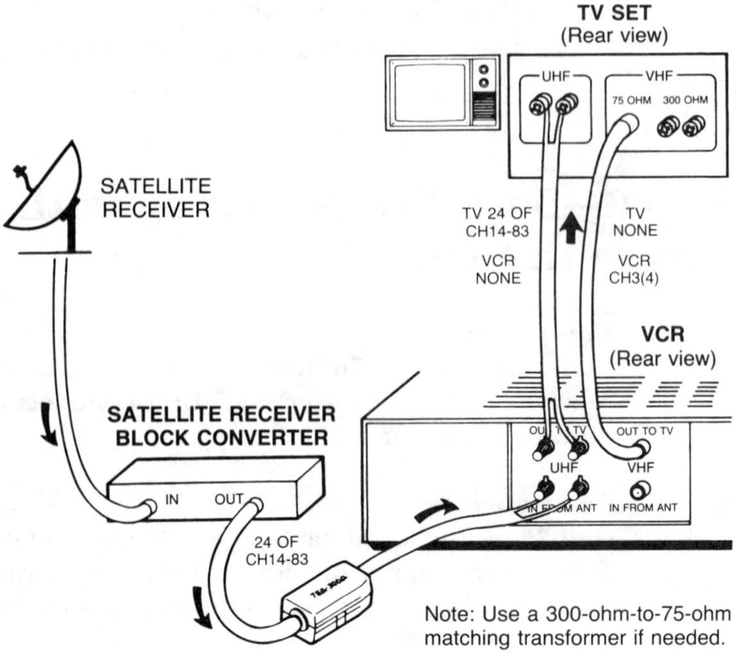

4. Connect the 300-ohm matching transformer output cable to the 300-ohm UHF IN terminals on the VCR.

To View a Prerecorded Tape

1. Be sure the VCR and TV power switches are on.
2. Set the TV/VCR switch on the VCR to VCR.
3. Tune the TV to the output channel of the VCR (channel 3 or 4).
4. Press the play button on the VCR.

To Record and View the Same Channel

1. Be sure all power switches are on.
2. Tune the VCR to the channel to record.
3. Set the TV/VCR switch on the VCR to VCR.
4. Tune the TV to the output channel of the VCR (channel 3 or 4).
5. View the TV to verify the channel to be recorded.
6. Press the record button(s) on the VCR.

To View the TV Without Recording

1. Turn the TV and satellite receiver power on. Turn the VCR power off.
2. Select the channel to view with the TV tuner.

To View One Channel While Recording Another Channel

1. Be sure all power switches are on.
2. Tune the VCR to the channel to record.
3. Set the TV/VCR switch on the VCR to VCR.
4. Tune the TV to the output channel of the VCR (channel 3 or 4).
5. View the TV to verify the channel to be recorded.
6. Press the record button(s) on the VCR.
7. Set the TV/VCR switch on the VCR to TV.
8. Tune the TV to the desired channel to view.

Hookup 7—Satellite Receiver with Down Converter to VCR

Equipment
Non-cable-ready VCR, non-cable-ready TV, satellite receiver, satellite-receiver down converter with VHF channel output on a 75-ohm coax.

Accessories
One 75-ohm coaxial cable. Place equipment in position and note cable length needed before purchasing cable.

Advantage
Channel operation is not confusing because only the satellite-receiver down converter can select the station.

Disadvantages
Cannot view one channel while recording another. VCR timer recording can only record one channel.

Hookup Procedure

Refer to figure 5-7 before hooking up equipment. If any equipment has already been installed, check each connection before making any changes. Some connections may not need to be changed.

1. Connect one end of a 75-ohm cable to the VHF OUT jack on the VCR. Connect the other end of the cable to the 75-ohm VHF input on the TV. *Note:* If you do not find a 75-ohm VHF antenna input on your TV, prepare the TV for a 75-ohm connection at the VHF input. See chapter 4.
2. Connect the 75-ohm VHF down converter output coax to the 75-ohm VHF IN jack on the VCR.

To View a Prerecorded Tape

1. Be sure the VCR and TV power switches are on and the satellite receiver is off.
2. Set the TV/VCR switch on the VCR to VCR.
3. Tune the TV to the output channel of the VCR (channel 3 or 4).
4. Press the play button on your VCR.

Figure 5-7
Hookup 7.

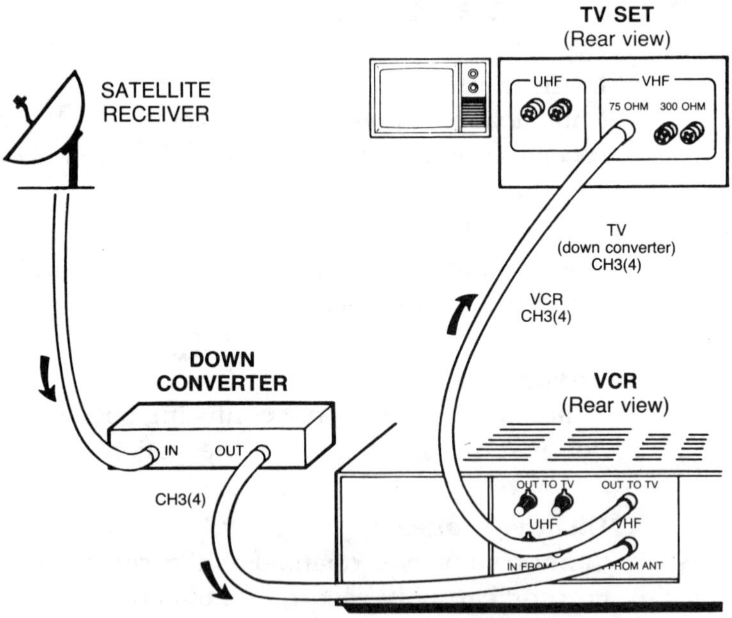

To View the TV Without Recording

1. Turn the TV and satellite receiver power on. Turn the VCR power off.
2. Set the TV channel to the down converter output (channel 3 or 4).
3. Select the channel to view with the down converter.

To Record and View the Same Channel

1. Be sure all power switches are on.
2. Tune the satellite down converter to the desired channel.
3. Tune the VCR to the down converter output channel. (Check your satellite receiver manual.)
4. Set the TV/VCR switch on the VCR to VCR.
5. Tune the TV to the output channel of the VCR (channel 3 or 4). *Note:* You may want to check the relative quality of the signal with the TV/VCR switch set to TV and the TV tuned to the output channel of the down converter.
6. View the TV to verify the channel to be recorded.
7. Press the record button(s) on the VCR.

Hookup 8—Videodisc Player Connection with Satellite Down Converter

Equipment
Non-cable-ready VCR, non-cable-ready TV, satellite receiver, satellite-receiver down converter with 75-ohm channel 3 (or 4) output cable, videodisc player.

Accessories
Three 75-ohm coaxial cables, one A/B signal switch. Place equipment in position and note cable lengths needed before purchasing cable.

Advantage
Can view a videodisc program while recording a satellite channel.

Disadvantage
Cannot record from a videodisc to the VCR.

Hookup Procedure

Refer to figure 5-8 before hooking up equipment. If any equipment has already been installed, check each connection before making any changes. Some connections may not need to be changed.

1. Connect the down converter channel 3 (or 4) output 75-ohm coax to the VHF IN jack on the VCR.
2. Connect one end of a 75-ohm cable to the VHF OUT jack on the VCR. Connect the other end of the coax to the B input on the signal switch.
3. Connect one end of a 75-ohm cable to the OUT jack on the videodisc player. Connect the other end of the coax to the A input on the signal switch.

Figure 5-8
Hookup 8.

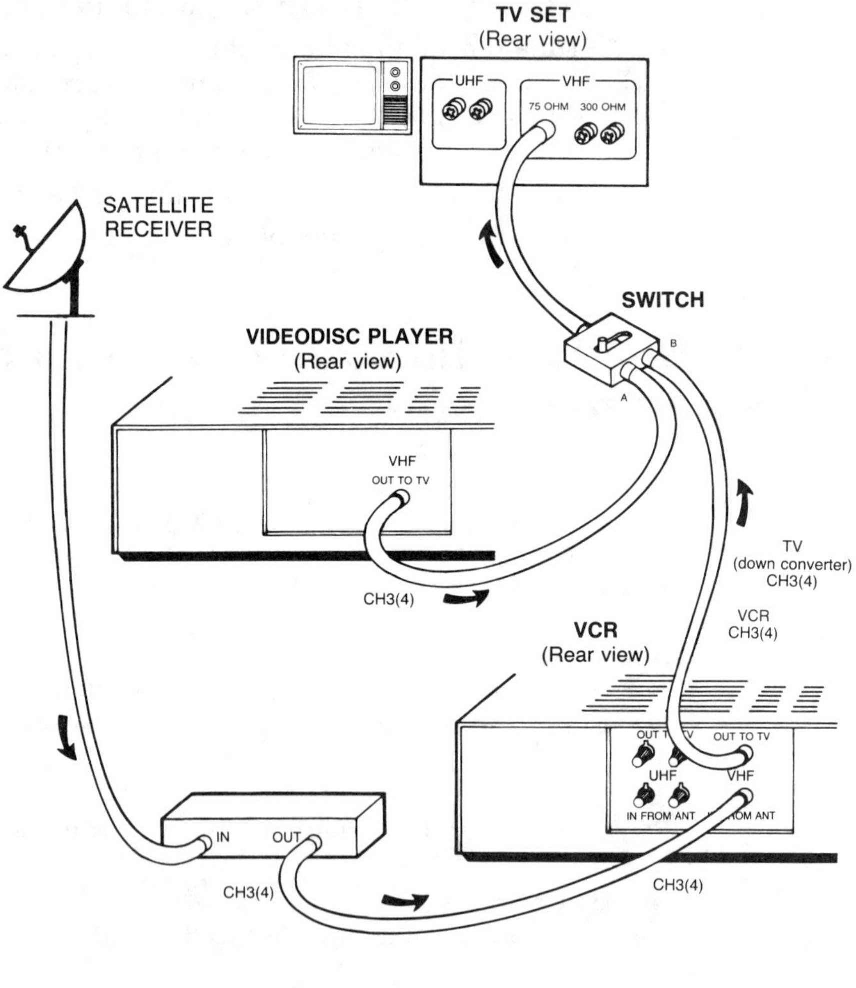

4. Connect one end of a 75-ohm cable to the OUT jack on the signal switch. Connect the other end of the coax to the 75-ohm VHF antenna input on the TV. *Note:* If you do not find a 75-ohm VHF antenna input on your TV, prepare the TV for a 75-ohm connection at the VHF input. See chapter 4.

To View a Prerecorded Tape

1. Be sure the TV and VCR power switches are on.
2. Set the TV/VCR switch on the VCR to VCR.
3. Tune the TV to the output channel of the VCR (channel 3 or 4).
4. Set the signal switch to B.
5. Press the play button on your VCR.

To View the TV Without Recording

1. Turn the TV and satellite receiver power on. Turn the VCR power off.
2. Set the TV channel to the down converter output (channel 3 or 4).
3. Select the channel to view with the down converter.

To Record and View a Satellite Channel

1. Be sure the satellite, VCR, and TV power switches are on.
2. Set the TV/VCR switch on the VCR to VCR.
3. Set the signal switch to B.
4. Tune the TV to the output channel of the VCR (channel 3 or 4).
5. Tune the down converter to the desired channel to record.
6. Tune the VCR to the down converter output (channel 3 or 4).
7. View the TV to verify the channel to record.
8. Press the record button(s) on the VCR.

To View a Videodisc While Recording a Satellite Channel

1. Be sure all power switches are on.
2. Set the TV/VCR switch on the VCR to VCR.
3. Set the signal switch to B.

4. Tune the TV to the output channel of the VCR (channel 3 or 4).
5. Tune the down converter to the desired channel to record.
6. Tune the VCR to the down converter output (channel 3 or 4).
7. View the TV to verify the channel to record.
8. Press the record button(s) on the VCR.
9. Set the signal switch to A.
10. Tune the TV to the output channel of the videodisc player (channel 3 or 4).
11. Press the play button on the videodisc player.

Hookup 9—Direct Cable TV Connection

Equipment
Cable-ready VCR, cable-ready TV, cable signal source.

Accessories
Two 75-ohm cables. Place equipment in position and note cable lengths needed before purchasing cable.

Advantages
Can record and view any unscrambled channel. Can use the programmable timer feature of the VCR for unscrambled channels.

Disadvantage
Cannot decode scrambled channels.

Hookup Procedure
Refer to figure 5-9 before hooking up equipment. If any equipment has already been installed, check each connection before making any changes. Some connections may not need to be changed.

1. Connect one end of a 75-ohm cable to the cable input. Connect the other end of the cable to the VHF IN jack on the VCR.
2. Connect one end of a 75-ohm cable to the VHF OUT jack on the VCR. Connect the other end of the cable to the 75-ohm VHF antenna input on the TV. *Note:* If you do not find a 75-

Figure 5-9
Hookup 9.

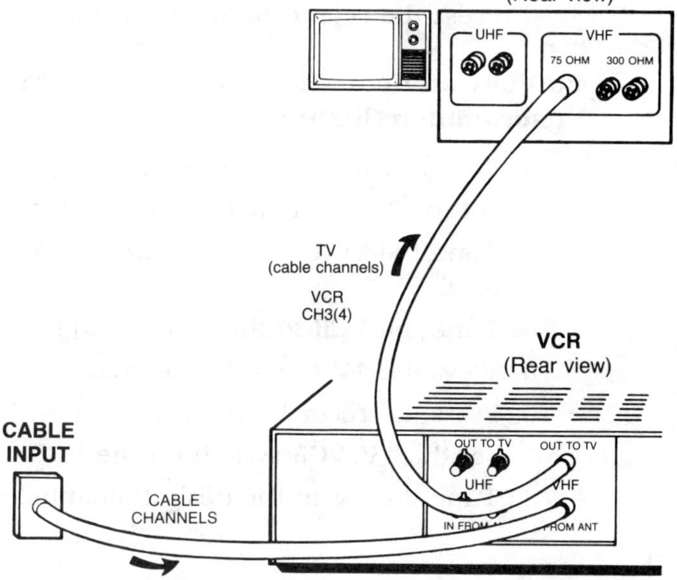

ohm VHF antenna input on your TV, prepare the TV for a 75-ohm connection at the VHF input. See chapter 4.

To View a Prerecorded Tape

1. Be sure the VCR and TV power switches are on.
2. Set the TV/VCR switch on the VCR to VCR.
3. Tune the TV to the output channel of the VCR.
4. Press the play button on the VCR.

To View the TV Without Recording

1. Turn the TV power on. Turn the VCR power off.
2. Select the channel to view with the TV tuner.

To Record and View an Unscrambled Channel

1. Be sure the power switches are on.
2. Set the TV/VCR switch on the VCR to VCR.
3. Tune the TV to the output channel of the VCR (channel 3 or 4).
4. Tune the VCR to the channel to record.

5. View the TV to verify the channel to be recorded.
6. Press the record button(s) on the VCR.

To View an Unscrambled Channel While Recording Another Unscrambled Channel

1. Be sure the power switches are on.
2. Set the TV/VCR switch on the VCR to VCR.
3. Tune the VCR to the output channel of the VCR (channel 3 or 4).
4. Tune the VCR to the channel to record.
5. View the TV to verify the channel to be recorded.
6. Press the record button(s) on the VCR.
7. Set the TV/VCR switch on the VCR to TV.
8. Tune the TV to the desired channel to view.

Hookup 10—Cable Converter-Decoder Connection

Equipment
Non-cable-ready VCR, non-cable-ready TV, converter-decoder, cable signal source.

Accessories
Three 75-ohm coaxial cables. Place equipment in position and note cable lengths needed before purchasing cable.

Advantages
Can record any unscrambled or decoded channel. No signal switch or splitter losses.

Disadvantages
Cannot watch another channel while recording. Cannot use the programmable timer feature of the VCR.

Hookup Procedure
Refer to figure 5-10 before hooking up equipment. If any equipment has already been installed, check each connection before making any changes. Some connections may not need to be changed.

Figure 5-10
Hookup 10.

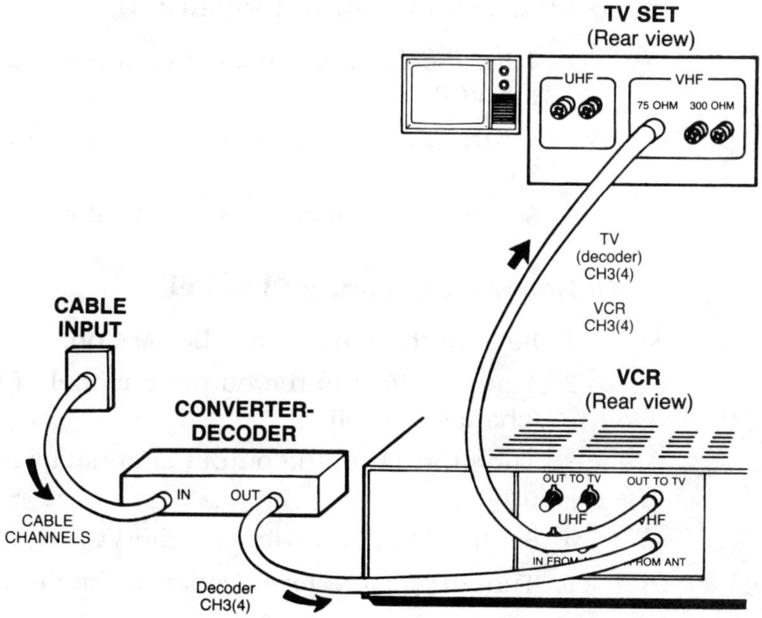

1. Connect one end of a 75-ohm cable to the cable input. Connect the other end of the coax to the converter-decoder input jack.
2. Connect one end of a 75-ohm cable to the converter-decoder output jack. Connect the other end of the cable to the VHF IN jack on the VCR.
3. Connect one end of a 75-ohm cable to the VHF OUT jack on the VCR. Connect the other end of the cable to the 75-ohm VHF antenna input jack on the TV. *Note:* If you do not find a 75-ohm VHF antenna input on your TV, prepare the TV for a 75-ohm connection at the VHF input. See chapter 4.

To View a Prerecorded Tape

1. Be sure the VCR and TV power switches are on and the cable box is off.
2. Set the TV/VCR switch on the VCR to VCR.
3. Tune the TV to the output channel of the VCR (channel 3 or 4).
4. Press the play button on the VCR.

To View the TV Without Recording

1. Turn the TV and cable converter-decoder power on. Turn the VCR power off.
2. Set the TV channel to the converter-decoder output (channel 3 or 4).
3. Select the channel to view with the cable converter-decoder.

To Record and View a Channel

1. Be sure the power switches are on.
2. Tune the VCR to the output channel of the converter-decoder (channel 3 or 4).
3. Tune the TV to the output channel of the VCR (channel 3 or 4).
4. Set the TV/VCR switch on the VCR to VCR.
5. Tune the converter-decoder to the desired channel to record.
6. View the TV to verify the channel to be recorded.
7. Press the record button(s) on the VCR.

To Check Alternative for Best Viewing

1. Be sure the power switches are on.
2. Tune the VCR to the output channel of the converter-decoder (channel 3 or 4).
3. Tune the TV to the output channel of the VCR (channel 3 or 4).
4. Set the TV/VCR switch on the VCR to VCR.
5. Tune the converter-decoder to the desired channel to record.
6. View the TV to verify the channel to be recorded.
7. Press the record button(s) on the VCR.
8. Tune the TV to the output channel of the converter-decoder (channel 3 or 4).
9. Set the TV/VCR switch on the VCR to TV or turn the VCR power off.
10. Note the relative signal quality when the switches are set as in steps 8 and 9 compared to signal quality when switches are set as in steps 1 through 6.

Important Notice

If during your hookup you must disconnect the incoming cable from the converter-decoder box, all or some of your cable channels may be disabled. Some cable TV systems sense the disconnection and turn off your cable service. After you have reconnected the cable, the signals are usually restored after a period of time. To have your service restored immediately, call your cable service company.

6 More Complex Hookups

All hookups in this chapter use a cable converter-decoder box, a TV set, and a VCR. The hookup you choose depends on the cable-ready status of your TV set and VCR. Other variations include two input cables instead of one, two converter-decoder boxes, and two VCRs.

Hookup 11—Cable-Ready TV and VCR (Version A)

Equipment
Cable-ready VCR, cable-ready TV, converter-decoder, cable signal source.

Accessories
Seven 75-ohm coaxial cables, one 75-ohm three-way signal splitter, two 75-ohm A/B signal switches. Place equipment in position and note cable lengths needed before purchasing cable.

Advantages
Can record and view the same channel, or record one channel and view another unscrambled channel, or record an unscrambled channel and view a scrambled channel. Can use the programmable timer channel change feature of the VCR for unscrambled channels.

Disadvantages

Cannot use the programmable timer channel change feature of the VCR for scrambled channels. Cannot view a scrambled channel while recording an unscrambled channel.

Hookup Procedure

Refer to figure 6-1 before hooking up equipment. If any equipment has already been installed, check each connection before making any changes. Some connections may not need to be changed.

1. Connect the cable input to the input of the signal splitter.

Figure 6-1
Hookup 11.

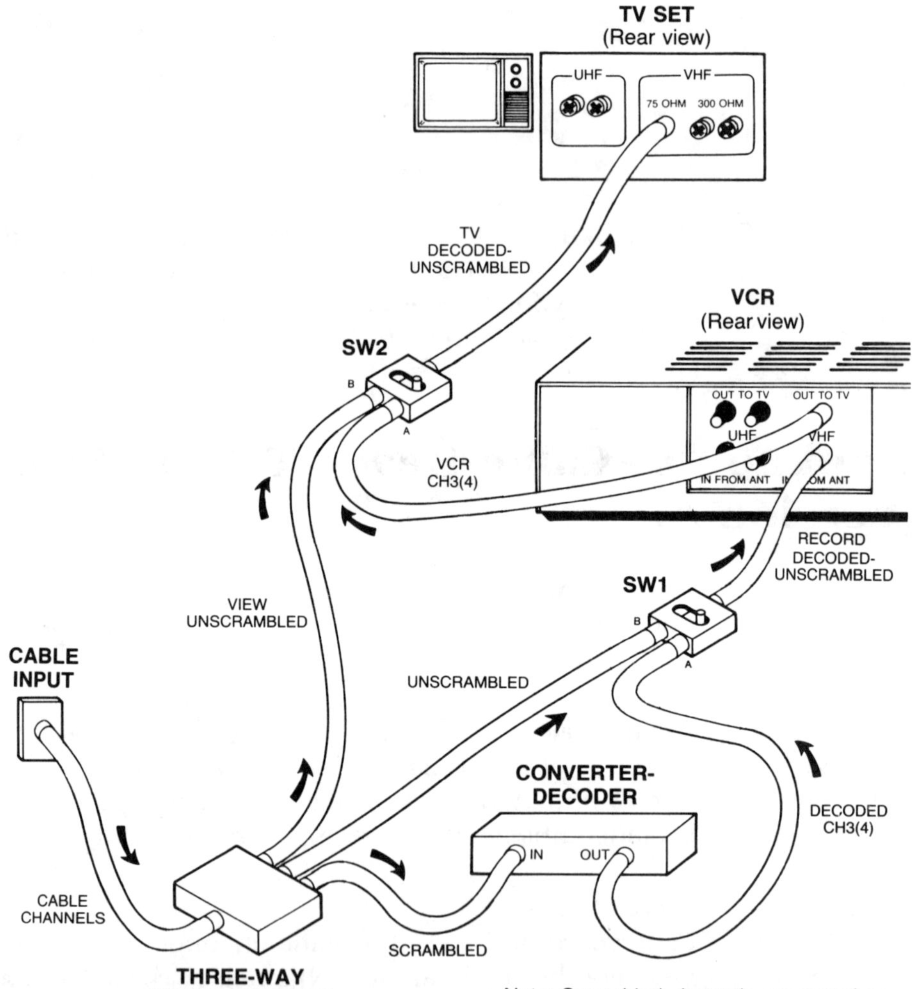

Note: Scrambled channels are routed on "unscrambled" cables and switches but are not viewable when tuned.

2. Connect one end of a 75-ohm cable to one output of the signal splitter. Connect the other end to the input of the converter-decoder.
3. Connect one end of a 75-ohm cable to the output on the converter-decoder. Connect the other end to the A input on switch SW1.
4. Connect one end of a 75-ohm cable to the second output on the signal splitter. Connect the other end to the B input on signal switch SW1.
5. Connect one end of a 75-ohm cable to the third output on the signal splitter. Connect the other end to the B input on switch SW2.
6. Connect one end of a 75-ohm cable to the VHF OUT jack of the VCR. Connect the other end to the A input on signal switch SW2.
7. Connect one end of a 75-ohm cable to the output jack on signal switch SW2. Connect the other end to the VHF IN on the TV set.

To View a Prerecorded Tape

1. Be sure the VCR and TV power switches are on and the cable converter is off.
2. Set SW2 to A.
3. Tune the TV to the output channel of the VCR (channel 3 or 4).
4. Set the TV/VCR switch on the VCR to VCR.
5. Press the play button on the VCR.

To Record and View the Same Scrambled or Unscrambled Channel

1. Set signal switch SW1 to A.
2. Set signal switch SW2 to A.
3. Be sure all power switches are on.
4. Tune the converter-decoder to the desired channel.
5. Tune the VCR to the converter-decoder output channel (channel 3 or 4).
6. Set the TV/VCR switch on the VCR to VCR.

7. Tune the TV to the VCR output channel (channel 3 or 4).
8. View the TV to verify the channel to be recorded.
9. Press the record button(s) on the VCR.

To View an Unscrambled Channel While Recording Another Unscrambled Channel

1. Set signal switch SW1 to B.
2. Set signal switch SW2 to A.
3. Set the TV/VCR switch on the VCR to VCR.
4. Be sure all power switches are on, except the converter-decoder.
5. Tune the VCR to the desired channel to record.
6. Tune the TV to the VCR output channel (channel 3 or 4).
7. View the TV to verify the channel to be recorded.
8. Press the record button(s) on the VCR.
9. Set the TV/VCR switch on the VCR to TV.
10. Tune the TV to the desired channel to view.

To View an Unscrambled Channel While Recording a Scrambled Channel

1. Set signal switch SW1 to A.
2. Set signal switch SW2 to B.
3. Set the TV/VCR switch on the VCR to VCR.
4. Be sure all power switches are on.
5. Tune the VCR to the converter-decoder output channel (channel 3 or 4).
6. Tune the converter-decoder to the desired scrambled channel.
7. Tune the TV to the VCR output channel (channel 3 or 4).
8. View the TV to verify the channel to be recorded.
9. Press the record button(s) on the VCR.
10. Set signal switch SW2 to B.
11. Tune the TV to the desired channel to view.

Special Notice

The hookups in this chapter are not only complex to connect but also require a knowledge of the system to set all A/B switches correctly. After you have decided on a hookup, make a table that indicates the position of each A/B switch for each desired operation. You may also want to designate the position of the TV/VCR switch on the VCR and what input channel each device is set to for each function. Refer to the operational details in each hookup for this data.

Hookup 12—Cable-Ready TV and VCR (Version B)

Equipment

Cable-ready VCR, cable-ready TV, converter-decoder, cable signal source.

Accessories

Twelve 75-ohm coaxial cables, two 75-ohm signal splitters (one in, two out), four 75-ohm A/B signal switches. Place equipment in position and note cable lengths needed before purchasing cable.

Advantages

Can record and view the same channel, or record one channel and view another unscrambled channel, or record an unscrambled channel and view a scrambled channel. Can use the programmable timer channel change feature of the VCR for unscrambled channels.

Disadvantage

Cannot use the programmable timer channel change feature of the VCR for scrambled channels.

Hookup Procedure

Refer to figure 6-2 before hooking up equipment. If any equipment has already been installed, check each connection before making any changes. Some connections may not need to be changed.

1. Connect one end of a 75-ohm cable to the cable input. Connect the other end to IN on signal splitter #1.

Figure 6-2
Hookup 12.

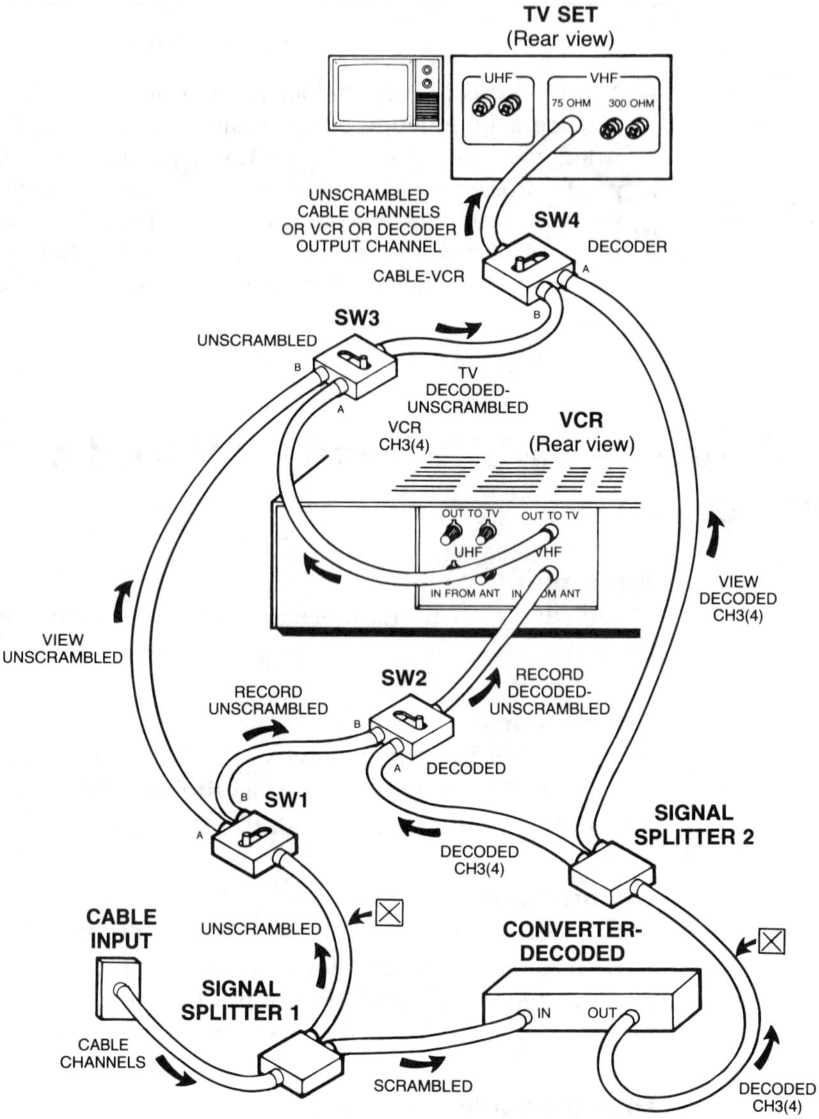

Notes: Scrambled channels are routed on "unscrambled" cables and switches but are not viewable when tuned. If signal strength is low, a VHF amplifier at ☒ will not risk overloading the cable converter-decoder.

2. Connect one end of a 75-ohm cable to one OUT of signal splitter #1. Connect the other end to IN on the converter-decoder.
3. Connect one end of a 75-ohm cable to OUT on the converter-decoder. Connect the other end to IN on signal splitter #2.

4. Connect one end of a 75-ohm cable to an OUT on signal splitter #2. Connect the other end to A on signal switch SW4.
5. Connect one end of a 75-ohm cable to IN on signal switch SW4. Connect the other end to the 75-ohm VHF antenna input on the TV. *Note:* If you do not find a 75-ohm VHF antenna input on your TV, prepare the TV for a 75-ohm connection at the VHF input. See chapter 4.
6. Test for signal strength.
   a. Turn on the TV and the converter-decoder.
   b. Set signal switch SW4 to A.
   c. Tune the TV to the output of the converter-decoder.
   d. Tune the converter-decoder to all stations you intend to view.
   e. If the signal is not satisfactory, choose a simple connection (such as hookup 10) or contact a local qualified technician. The cable is not supplying enough signal for so many devices.
   f. If the signal is satisfactory, continue with this hookup.
7. Connect one end of a 75-ohm cable to the other OUT of signal splitter #1. Connect the other end to IN on signal switch SW1.
8. Connect one end of a 75-ohm cable to the other OUT of signal splitter #2. Connect the other end to A on signal switch SW2.
9. Connect one end of a 75-ohm cable to B on signal SW1. Connect the other end to B on signal SW2.
10. Connect one end of 75-ohm cable to IN on signal SW2. Connect the other end to VHF IN on the VCR.
11. Connect one end of 75-ohm cable to VHF OUT on the VCR. Connect the other end to A on signal SW3.
12. Connect one end of a 75-ohm cable to B on signal SW3. Connect the other end to A on signal SW1.
13. Connect one end of a 75-ohm cable to IN on signal SW3. Connect the other end to B on signal SW4.

To View a Prerecorded Tape

1. Be sure the VCR and TV power switches are on.
2. Set SW3 to A.

3. Set SW4 to B.
4. Tune the TV to the output channel of the VCR (channel 3 or 4).
5. Set the TV/VCR switch on the VCR to VCR.
6. Press the play button on the VCR.

To Record and View the Same Scrambled or Unscrambled Channel

You may want to compare the reception quality of the alternative path. In the following steps, change the TV/VCR switch on the VCR to TV and tune the TV to the output channel of the converter-decoder.

1. Set signal switch SW2 to A.
2. Set signal switch SW3 to A.
3. Set signal switch SW4 to B.
4. Be sure all power switches are on.
5. Tune the converter-decoder to the desired channel.
6. Tune the VCR to the converter-decoder output channel (channel 3 or 4).
7. Set the TV/VCR switch on the VCR to VCR.
8. Tune the TV to the VCR output channel (channel 3 or 4).
9. View the TV to verify the channel to be recorded.
10. Press the record button(s) on the VCR.

To View an Unscrambled Channel While Recording Another Unscrambled Channel

You may want to compare the reception quality of the alternative path. In the following steps, turn on the converter-decoder, set signal SW4 to A, and tune the TV to the output channel of the converter-decoder.

1. Set signal switch SW1 to B.
2. Set signal switch SW2 to B.
3. Set signal switch SW3 to A.
4. Set signal switch SW4 to B.
5. Set the TV/VCR switch on the VCR to VCR.
6. Be sure all power switches are on, except the converter-decoder.
7. Tune the VCR to the desired channel to record.

8. Tune the TV to the VCR output channel (channel 3 or 4).
9. View the TV to verify the channel to be recorded.
10. Press the record button(s) on the VCR.
11. Set the TV/VCR switch on the VCR to TV.
12. Tune the TV to the desired channel to view.

To View an Unscrambled Channel While Recording a Scrambled Channel

1. Set signal switch SW1 to A.
2. Set signal switch SW2 to A.
3. Set signal switch SW3 to A.
4. Set signal switch SW4 to B.
5. Set the TV/VCR switch on the VCR to VCR.
6. Be sure all power switches are on.
7. Tune the VCR to the converter-decoder output channel (channel 3 or 4).
8. Tune the converter-decoder to the desired scrambled channel.
9. Tune the TV to the VCR output channel (channel 3 or 4).
10. View the TV to verify the channel to be recorded.
11. Press the record button(s) on the VCR.
12. Set signal switch SW3 to B.
13. Tune the TV to the desired channel to view.

To View a Scrambled Channel While Recording an Unscrambled Channel

1. Set signal switch SW1 to B.
2. Set signal switch SW2 to B.
3. Set signal switch SW3 to A.
4. Set signal switch SW4 to B.
5. Set the TV/VCR switch on the VCR to VCR.
6. Be sure all power switches are on.
7. Tune the TV to the VCR output channel (channel 3 or 4).
8. View the TV to verify the channel to be recorded.
9. Press the record button(s) on the VCR.

10. Set signal switch SW4 to A.
11. Tune the TV to the converter-decoder output channel (channel 3 or 4).
12. Tune the converter-decoder to the desired channel to view.

Hookup 13—Cable-Ready TV, Non-Cable-Ready VCR

Equipment
Non-cable-ready VCR, cable-ready TV, converter-decoder.

Accessories
Six 75-ohm coaxial cables, one 75-ohm signal splitter (one in, two out), one 75-ohm A/B signal switch. Place equipment in position and note cable lengths needed before purchasing cable.

Advantage
Can view any unscrambled channel while recording any channel.

Disadvantages
Cannot view scrambled channel while recording unscrambled channel. Cannot use the programmable timer channel change feature of the VCR.

Hookup Procedure
Refer to figure 6-3 before hooking up equipment. If any equipment has already been installed, check each connection before making any changes. Some connections may not need to be changed.

1. Connect one end of a 75-ohm cable to the cable input. Connect the other end to IN on the signal splitter.
2. Connect one end of a 75-ohm cable to one OUT of the signal splitter. Connect the other end to IN on the converter-decoder.
3. Connect one end of a 75-ohm cable to OUT on the converter-decoder. Connect the other end to VHF IN on the VCR.
4. Connect one end of a 75-ohm cable to the other OUT of the signal splitter. Connect the other end to A on the signal switch.

Figure 6-3
Hookup 13.

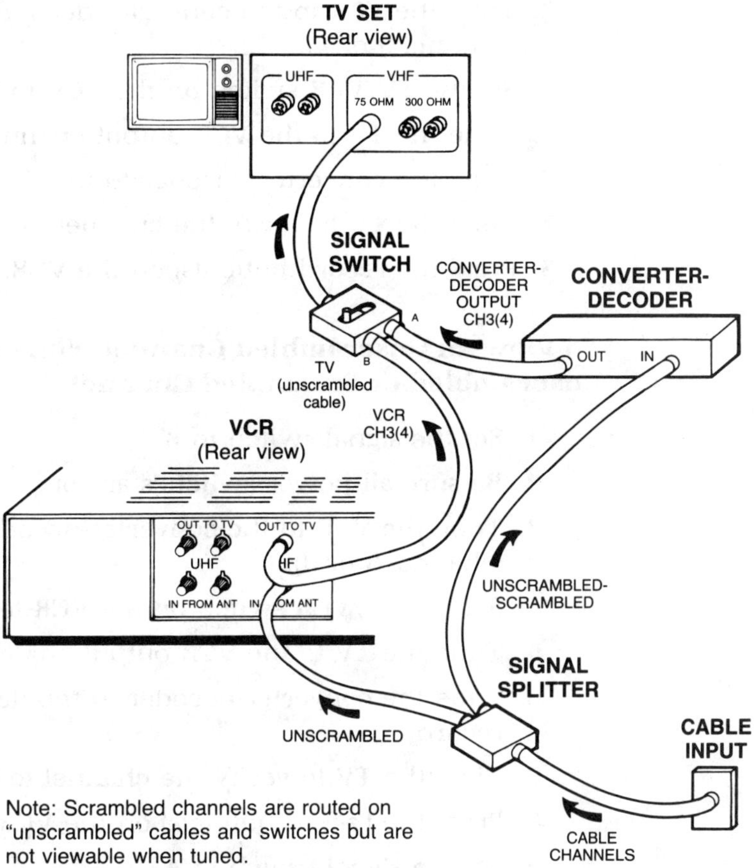

5. Connect one end of 75-ohm cable to VHF OUT on the VCR. Connect the other end to B on the signal switch.
6. Connect one end of a 75-ohm cable to OUT on the signal switch. Connect the other end to the 75-ohm VHF antenna input on the TV. *Note:* If you do not find a 75-ohm VHF antenna input on your TV, prepare the TV for a 75-ohm connection at the VHF input. See chapter 4.

To Record and View a Scrambled or Unscrambled Channel

You may want to compare the reception quality of the alternative path. In the following steps, change the TV/VCR switch on the VCR to TV and tune the TV to the output channel of the converter-decoder.

1. Set the signal switch to B.
2. Be sure all power switches are on.

3. Tune the VCR to the converter-decoder output channel (channel 3 or 4).
4. Set the TV/VCR switch on the VCR to VCR.
5. Tune the TV to the VCR output channel (channel 3 or 4).
6. Tune the converter-decoder to the desired channel to record.
7. View the TV to verify the channel to be recorded.
8. Press the record button(s) on the VCR.

To View an Unscrambled Channel While Recording a Different Unscrambled or Scrambled Channel

1. Set the signal switch to B.
2. Be sure all power switches are on.
3. Tune the VCR to the converter-decoder output channel (channel 3 or 4).
4. Set the TV/VCR switch on the VCR to VCR.
5. Tune the TV to the VCR output channel (channel 3 or 4).
6. Tune the converter-decoder to the desired channel to record.
7. View the TV to verify the channel to be recorded.
8. Press the record button(s) on the VCR.
9. Set the signal switch to A.
10. Tune the TV to the desired channel to view.

Hookup 14—Cable-Ready VCR, Non-Cable-Ready TV (Version A)

Equipment

Cable-ready VCR, non-cable-ready TV, converter-decoder, cable signal source.

Accessories

Six 75-ohm coaxial cables, one 75-ohm signal splitter (one in, two out), one 75-ohm A/B signal switch. Place equipment in position and note cable lengths needed before purchasing cable.

Advantages

Can view any channel while recording any unscrambled channel. Can use the programmable timer channel change feature of the VCR for unscrambled channels.

Disadvantage

Cannot record scrambled channels.

Hookup Procedure

Refer to figure 6-4 before hooking up equipment. If any equipment has already been installed, check each connection before making any changes. Some connections may not need to be changed.

1. Connect the cable input to the input on the signal splitter.

Figure 6-4
Hookup 14.

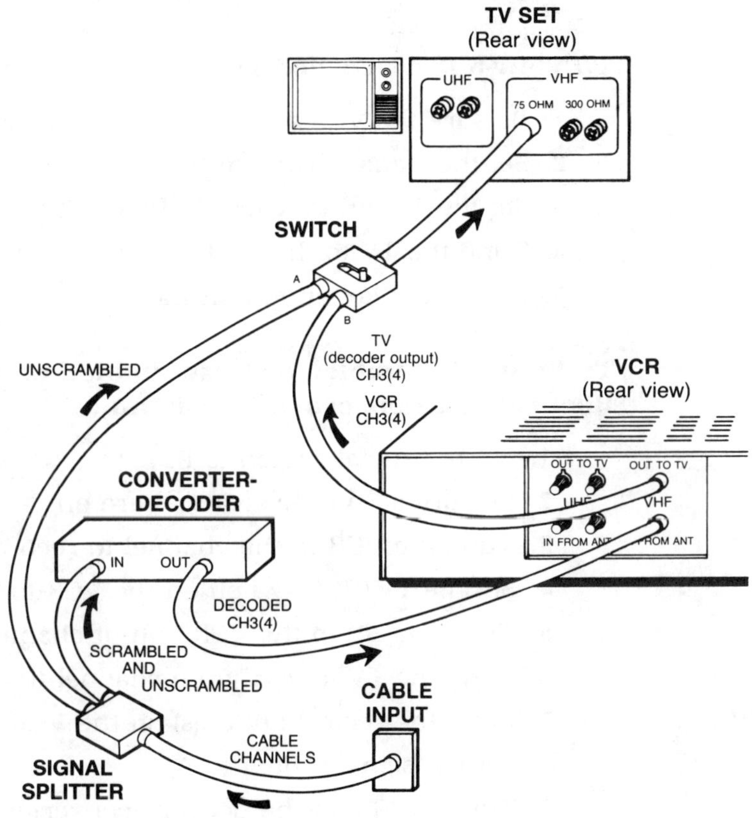

Note: Scrambled channels are routed on "unscrambled" cables and switches but are not viewable when tuned.

2. Connect one end of a 75-ohm cable to one output of the signal splitter. Connect the other end of this cable to the input of the converter-decoder.
3. Connect one end of a 75-ohm cable to the output of the converter-decoder. Connect the other end to A on the signal switch.
4. Connect one end of a 75-ohm cable to the other output of the signal splitter. Connect the other end to VHF IN on the VCR.
5. Connect one end of a 75-ohm cable to VHF OUT on the VCR. Connect the other end to B on the signal switch.
6. Connect one end of a 75-ohm cable to the output on the signal switch. Connect the other end to the 75-ohm VHF antenna input on the TV. *Note:* If you do not find a 75-ohm VHF antenna input on your TV, prepare the TV for a 75-ohm connection at the VHF input. See chapter 4.

To View a Prerecorded Tape

1. Be sure the VCR and TV power switches are on.
2. Set the signal switch to B.
3. Set the TV/VCR switch on the VCR to VCR.
4. Tune the TV to the VCR output channel (channel 3 or 4).
5. Press the play button on the VCR.

To View a Scrambled or Unscrambled Channel While Recording an Unscrambled Channel

1. Set the signal switch to B.
2. Be sure all power switches are on.
3. Tune the VCR to the channel to record.
4. Set the TV/VCR switch on the VCR to VCR.
5. Tune the TV to the VCR output channel (channel 3 or 4).
6. View the TV to verify the channel to be recorded.
7. Press the record button(s) on the VCR.
8. Set the signal switch to A.
9. Tune the TV to the decoder-converter output channel (channel 3 or 4).
10. Tune the decoder-converter to the desired viewing channel.

Hookup 15—Cable-Ready VCR, Non-Cable-Ready TV (Version B, Two-Cable Signal Source)

Equipment
Cable-ready VCR, non-cable-ready TV, two-cable input converter-decoder, two-cable signal source.

Accessories
Ten 75-ohm coaxial cables, two 75-ohm signal splitters (one in, two out), two 75-ohm A/B signal switches. Place equipment in position and note cable lengths needed before purchasing cable.

Advantages
Can view any channel while recording any unscrambled channel. Can use the programmable timer channel change feature of the VCR for channels on one cable. (The VCR cannot select the other cable.)

Disadvantages
Cannot record scrambled channels. Cannot use the programmable timer channel change feature of the VCR on alternate cable channels.

Hookup Procedure
Refer to figure 6-5 before hooking up equipment. *Note:* If any equipment has already been installed, check each connection before making any changes. Some connections may not need to be changed.

1. Connect one end of a 75-ohm cable to cable input A. Connect the other end to IN on signal splitter #1.
2. Connect one end of a 75-ohm cable to cable input B. Connect the other end to IN on signal splitter #2.
3. Connect one end of a 75-ohm cable to one OUT of signal splitter #1. Connect the other end to IN A on the converter-decoder.
4. Connect one end of a 75-ohm cable to one OUT on signal splitter #2. Connect the other end to IN B on the converter-decoder.
5. Connect one end of a 75-ohm cable to OUT on the converter-decoder. Connect the other end to B on signal switch SW2.

Figure 6-5
Hookup 15.

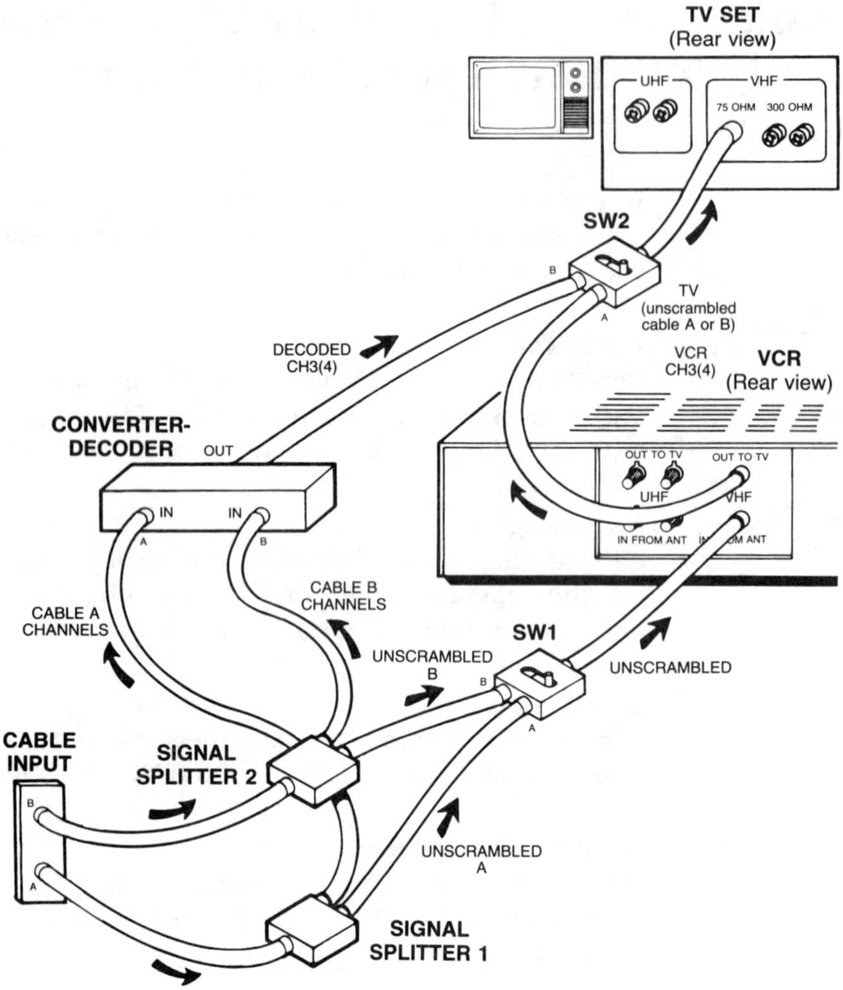

6. Connect one end of a 75-ohm cable to the other OUT of signal splitter #1. Connect the other end to A on signal switch SW1.
7. Connect one end of a 75-ohm cable to OUT on signal switch SW1. Connect the other end to VHF IN on the VCR.
8. Connect one end of a 75-ohm cable to VHF OUT on the VCR. Connect the other end to A on signal switch SW2.
9. Connect one end of a 75-ohm cable to OUT on signal switch SW2. Connect the other end to the 75-ohm VHF antenna

input on the TV. *Note:* If you do not find a 75-ohm VHF antenna input on your TV, you will have to prepare the TV for a 75-ohm connection at the VHF input. See chapter 4.

To View a Prerecorded Tape

1. Be sure the TV and VCR power switches are on.
2. Set signal switch SW2 to A.
3. Set the VCR/TV switch on the VCR to VCR.
4. Tune the TV to the VCR output channel (channel 3 or 4).
5. Press the play button on the VCR.

To Record and View an Unscrambled Channel

1. Be sure the TV and VCR power switches are on.
2. Set signal switch SW2 to A.
3. Set the TV/VCR switch on the VCR to VCR.
4. Tune the TV to the output channel of the VCR (channel 3 or 4).
5. Select the channel to record by setting signal switch SW1 and tuning the VCR.
6. View the TV to verify the channel to be recorded.
7. Press the record button(s) on the VCR.

To View an Unscrambled or Scrambled Channel While Recording an Unscrambled Channel

1. Be sure the power switches are on.
2. Set signal switch SW2 to A.
3. Set the TV/VCR switch on the VCR to VCR.
4. Tune the TV to the output channel of the VCR (channel 3 or 4).
5. Select the channel to record by tuning the VCR.
6. Press the record button(s) on the VCR.
7. View the TV to verify that the correct channel is recording.
8. Set signal switch SW2 to B.
9. Set the TV set channel to the output of the converter-decoder (channel 3 or 4).
10. Tune the converter-decoder to the desired channel to view.

Hookup 16—Cable-Ready VCR, Non-Cable-Ready TV (Version C)

Equipment
Cable-ready VCR, non-cable-ready TV, converter-decoder, cable signal source.

Accessories
Nine 75-ohm coaxial cables, three 75-ohm A/B signal switches, one 75-ohm signal splitter (one in, two out). Place equipment in position and note cable lengths needed before purchasing cable.

Advantages
Can record any unscrambled or decoded channel. Can use the programmable timer channel change feature of the VCR for unscrambled channels.

Disadvantages
Cannot watch another channel while recording a decoded channel. Cannot use the programmable timer channel change feature of the VCR for scrambled channels.

Hookup Procedure
Refer to figure 6-6 before hooking up equipment. If any equipment has already been installed, check each connection before making any changes. Some connections may not need to be changed.

1. Connect one end of a 75-ohm cable to the cable input. Connect the other end to IN of the signal splitter.
2. Connect one end of a 75-ohm cable to one OUT of the signal splitter. Connect the other end to IN on the converter-decoder.
3. Connect one end of a 75-ohm cable to OUT on the converter-decoder. Connect the other end to OUT on signal SW2.
4. Connect one end of a 75-ohm cable to B on signal SW2. Connect the other end to A on signal SW3.
5. Connect one end of a 75-ohm cable to IN on signal SW3. Connect the other end to the 75-ohm VHF antenna input on the TV. *Note:* If you do not find a 75-ohm VHF antenna input on your TV, prepare the TV for a 75-ohm connection at the VHF input. See chapter 4.

Figure 6-6
Hookup 16.

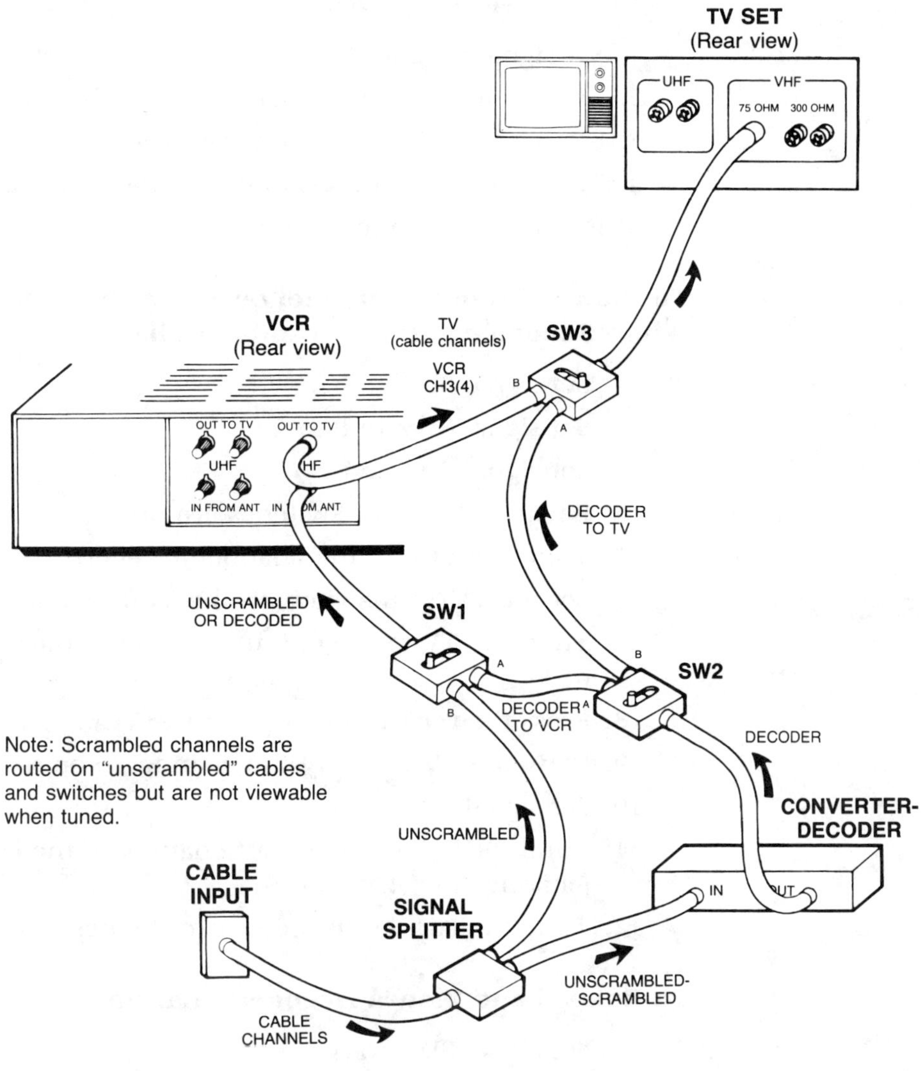

6. Connect one end of a 75-ohm cable to A on signal SW2. Connect the other end to A on signal SW1.
7. Connect one end of a 75-ohm cable to OUT on signal SW1. Connect the other end to VHF IN on the VCR.
8. Connect one end of a 75-ohm cable to VHF OUT on the VCR. Connect the other end to B on signal SW3.
9. Connect one end of a 75-ohm cable to the other OUT on the signal splitter. Connect the other end to B on signal SW1.

To View a Prerecorded Tape

1. Set signal SW3 to B.
2. Be sure the VCR and TV power switches are on.
3. Set the VCR/TV switch on the VCR to VCR.
4. Tune the TV to the VCR output channel (channel 3 or 4).
5. Press the play button on the VCR.

To View a Decoded Channel or an Unscrambled Channel While Recording an Unscrambled Channel

1. Set signal SW1 to B.
2. Set signal SW2 to B.
3. Set signal SW3 to B.
4. Be sure all power switches are on.
5. Tune the VCR to the channel to record.
6. Set the TV/VCR switch on the VCR to VCR.
7. Tune the TV to the output channel of the VCR (channel 3 or 4).
8. Press the record button(s) on the VCR.
9. View the TV to verify the channel to be recorded.
10. Set signal SW3 to A.
11. Tune the TV to the output channel of the converter-decoder (channel 3 or 4).
12. Tune the converter-decoder to the desired channel to view.

To Record and View a Decoded Channel

1. Set signal SW1 to A.
2. Set signal SW2 to A.
3. Set signal SW3 to B.
4. Be sure all power switches are on.
5. Tune the VCR to the output channel of the converter-decoder (channel 3 or 4).
6. Tune the converter-decoder to the desired channel to view.
7. Set the TV/VCR switch on the VCR to VCR.
8. Tune the TV to the output channel of the VCR (channel 3 or 4).

9. Press the record button(s) on the VCR.
10. View the TV to verify the channel to record and view.

To Check Alternative for Best Operation

1. Set signal SW1 to A.
2. Set signal SW2 to A.
3. Set signal SW3 to B.
4. Be sure all power switches are on.
5. Tune the VCR to the output channel of the converter-decoder (channel 3 or 4).
6. Tune the converter-decoder to the desired channel to view.
7. Set the TV/VCR switch on the VCR to VCR.
8. Tune the TV to the output channel of the VCR (channel 3 or 4).
9. View the TV to verify the channel to record and view.
10. Tune the TV to the output channel of the converter-decoder (channel 3 or 4).
11. Set the TV/VCR switch on the VCR to TV.
12. Note relative signal quality when using settings in steps 7 and 8 compared to quality when using settings in steps 10 and 11.

Hookup 17—Non-Cable-Ready TV and VCR

Equipment
Non-cable-ready VCR, non-cable-ready TV, converter-decoder, cable-to-UHF block converter with 300-ohm twin lead output or matching transformer.

Accessories
Five 75-ohm coaxial cables, one 300-ohm twin lead cable, one 75-ohm signal splitter (one in, two out). Place equipment in position and note cable lengths needed before purchasing cable.

Advantages
Can record any unscrambled channel while viewing a decoded channel. Can view any unscrambled channel while recording a decoded channel.

Disadvantage

Cannot use the programmable timer channel change feature of the VCR to select an alternate scrambled channel.

Hookup Procedure

Refer to figure 6-7 before hooking up equipment. If any equipment has already been installed, check each connection before making any changes. Some connections may not need to be changed.

1. Connect one end of a 75-ohm cable to VHF OUT on the VCR. Connect the other end to the 75-ohm VHF antenna input on

Figure 6-7
Hookup 17.

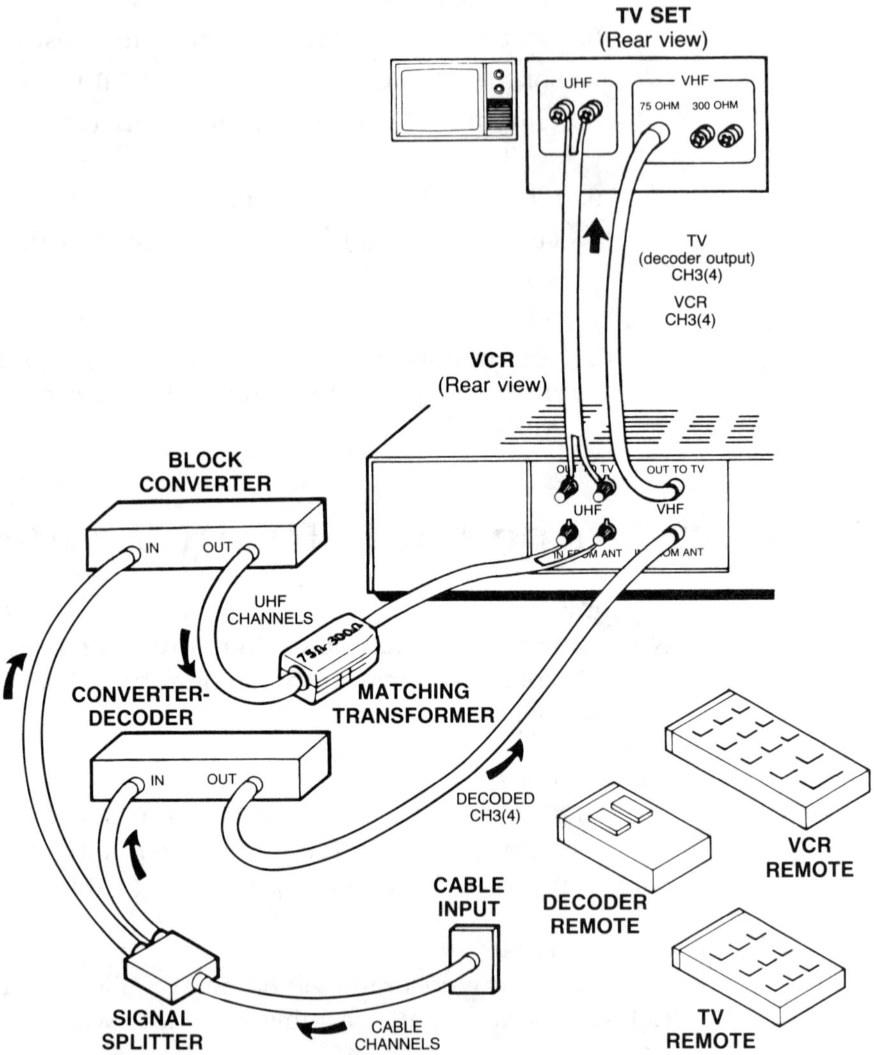

the TV. *Note:* If you do not find a 75-ohm VHF antenna input on your TV, you will have to prepare the TV for a 75-ohm connection at the VHF input. See chapter 4.

2. Connect one end of the 300-ohm twin lead to UHF OUT on the VCR. Connect the other end to the 300-ohm UHF antenna input on the TV.
3. Connect the 300-ohm twin lead of the UHF output of the block converter to UHF IN on the VCR.
4. Connect one end of a 75-ohm cable to the output of the converter-decoder. Connect the other end to VHF IN on the VCR.
5. Connect one end of a 75-ohm cable to the input of the converter-decoder. Connect the other end to an output of the signal splitter.
6. Connect one end of a 75-ohm cable to the input of the block converter. Connect the other end to the other output of the signal splitter.
7. Connect one end of a 75-ohm cable to the cable input. Connect the other end to the input of the signal splitter.

To View a Prerecorded Tape

1. Be sure the TV and VCR power switches are on.
2. Set the VCR/TV switch on the VCR to VCR.
3. Tune the TV to the output channel of the VCR (channel 3 or 4).
4. Press the play button on the VCR.

To Record and View the Same Unscrambled Channel

1. Be sure all power switches are on.
2. Tune the VCR to the channel to record.
3. Set the TV/VCR switch on the VCR to VCR.
4. Tune the TV to the output channel of the VCR (channel 3 or 4).
5. View the TV to verify the channel to be recorded.
6. Press the record button(s) on the VCR.

To View an Unscrambled Channel While Recording Another Unscrambled Channel

1. Be sure all power switches are on.
2. Tune the VCR to the channel to record.
3. Set the TV/VCR switch on the VCR to VCR.
4. Tune the TV to the output channel of the VCR (channel 3 or 4).
5. View the TV to verify the channel to be recorded.
6. Press the record button(s) on the VCR.
7. Set the TV/VCR switch on the VCR to TV.
8. Tune the TV to the desired channel to view.

To Record and View the Same Scrambled Channel

1. Be sure all power switches are on.
2. Tune the VCR to the output channel of the converter-decoder (channel 3 or 4).
3. Set the TV/VCR switch on the VCR to VCR.
4. Tune the TV to the output channel of the VCR (channel 3 or 4).
5. Tune the converter-decoder to the desired channel to record.
6. View the TV to verify the channel to be recorded.
7. Press the record button(s) on the VCR.

To View an Unscrambled Channel While Recording a Scrambled Channel

1. Be sure all power switches are on.
2. Tune the VCR to the output channel of the converter-decoder.
3. Set the TV/VCR switch on the VCR to VCR.
4. Tune the TV to the output channel of the VCR (channel 3 or 4).
5. Tune the converter-decoder to the desired channel to record.
6. View the TV to verify the channel to be recorded.
7. Press the record button(s) on the VCR.
8. Set the TV/VCR switch on the VCR to TV.
9. Tune the TV to the desired channel to view.

To View a Scrambled Channel While Recording an Unscrambled Channel

1. Be sure all power switches are on.
2. Tune the TV to the VCR output channel (channel 3 or 4).
3. Tune the VCR to the channel to record.
4. Set the TV/VCR switch on the VCR to VCR.
5. Press the record button(s) on the VCR.
6. View the TV to verify the channel to record.
7. Set the TV/VCR switch on the VCR to TV.
8. Tune the TV to the converter-decoder output channel.
9. Tune the converter-decoder to the channel to view.

Hookup 18—Two Cable Converter-Decoder Boxes

Equipment
TV set (cable-ready or non-cable-ready), VCR (cable-ready or non-cable-ready), two converter-decoder boxes.

Accessories
One two-way signal splitter, one A/B signal switch, six 75-ohm cables with F-plugs. Place equipment in position and note cable lengths needed before purchasing cables.

Advantages
Can record one scrambled premium channel while viewing another scrambled premium channel. Can use the programmable timer channel change feature of the VCR for scrambled channels.

Disadvantages
None.

Hookup Procedure
Refer to figure 6-8 before hooking up equipment. If any equipment has already been installed, check each connection before making any changes. Some connections may not need to be changed.

Figure 6-8
Hookup 18.

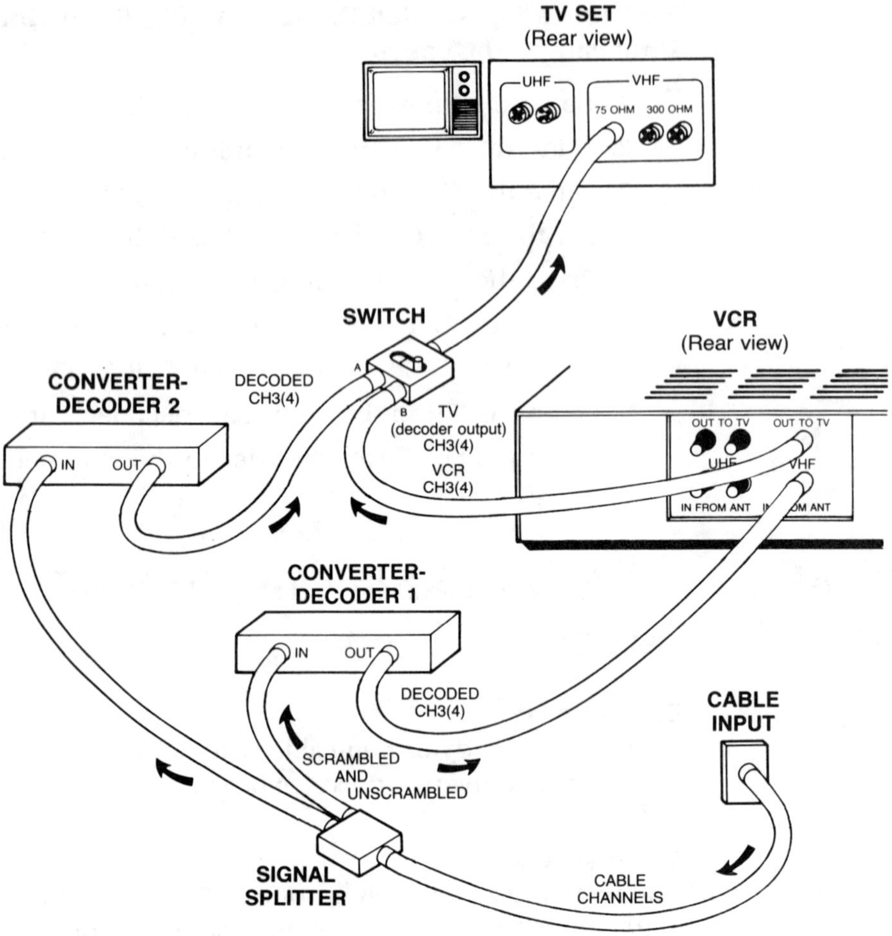

1. Connect the input cable to the input jack of a two-way signal splitter.
2. Connect one end of a 75-ohm cable to one output of the signal splitter. Connect the other end to the input jack on converter-decoder #1.
3. Connect one end of a 75-ohm cable to the other output of the signal splitter. Connect the other end to the input jack on converter-decoder #2.
4. Connect one end of 75-ohm cable to the output jack of converter-decoder #1. Connect the other end to the VHF IN jack on the VCR.

5. Connect one end of a 75-ohm cable to the VHF OUT jack on the VCR. Connect the other end to the B input of a signal switch.
6. Connect one end of a 75-ohm cable to the output jack on converter-decoder #2. Connect the other end to the A input on the signal switch.
7. Connect one end of a 75-ohm cable to the output jack of the signal switch. Connect the other end to the VHF antenna jack on the TV set.

To View a Prerecorded Tape

1. Be sure the VCR and TV power switches are on. Turn off both converter-decoder boxes.
2. Set the TV/VCR switch on the VCR to VCR.
3. Tune the TV to the output channel of the VCR (channel 3 or 4).
4. Set the signal switch to B.
5. Press the play button on the VCR.

To Record and View the Same Channel

1. Be sure the TV and VCR power switches are on. Turn on converter-decoder #1. Turn off converter-decoder #2.
2. Tune the VCR to the output channel of the converter-decoder (channel 3 or 4).
3. Tune the TV to the output channel of the VCR (channel 3 or 4).
4. Set the signal switch to B.
5. Set the TV/VCR switch on the VCR to VCR.
6. Tune the converter-decoder to the desired channel with converter-decoder #1.
7. Press the record button(s) on the VCR.

To View a Channel Without Recording

1. Turn on the TV set and converter-decoder #1. Turn off the VCR and converter-decoder #2.
2. Set the signal switch to B.

3. Tune the TV to the output channel of converter-decoder #1.
4. Select the channel to view with converter-decoder #1.

To View Any Scrambled or Unscrambled Channel While Recording Any Other Scrambled or Unscrambled Channel

1. Turn on power to all units.
2. Set the signal switch to B.
3. Tune the VCR to the output channel of converter-decoder #1 (channel 3 or 4).
4. Tune converter-decoder #1 to the channel to be recorded.
5. Set the TV/VCR switch on the VCR to TV.
6. View the TV to verify the channel to record.
7. Press the record button(s) on the VCR.
8. Set the signal switch to A.
9. Tune the TV set to the output channel of converter-decoder #2 (channel 3 or 4).
10. Tune the channel to be viewed with converter-decoder #2.

Hookup 19—Two VCRs

Equipment
Two VCRs (cable-ready), cable-ready TV set, cable converter-decoder, two cable signal sources.

Accessories
Six 75-ohm coax cables with F-plugs, two 75-ohm coax cables with RCA phono plugs, one signal splitter (two-to-one), one three-way A/B/C signal switch. Place equipment in position and note cable lengths needed before purchasing cables.

Advantages
Can record or view any scrambled or unscrambled channel simultaneously in any combination. Can copy video tapes. Can record any scrambled or unscrambled channel while viewing a prerecorded tape.

Disadvantages
None.

Hookup Procedure

Refer to figure 6-9 before hooking up equipment. Some converter-decoder boxes come with two input connectors, one for the cable carrying the unscrambled channels and the other for the cable carrying the scrambled premium channels. Usually, both cables connect to the connectors on the converter-decoder box. In this hookup, only the scrambled cable is connected so that the converter-decoder box can unscramble premium channels. Unscrambled channels are selected with the cable-ready TV set or VCR.

1. Connect the scrambled signal cable to the scrambled input on the converter-decoder box.
2. Connect the unscrambled cable to the signal splitter input.
3. Connect a 75-ohm cable from one output of the signal splitter to the A input on the A/B/C switch.
4. Connect a 75-ohm cable from the other output of the signal splitter to the VHF input on VCR 2.
5. Connect a 75-ohm cable from the output of the converter-decoder box to the VHF input on VCR 1.
6. Connect a 75-ohm coax cable between the VHF output of VCR 1 and the B input of the A/B/C signal switch.
7. Connect a 75-ohm cable between the VHF output of VCR 2 and the C input of the A/B/C signal switch.
8. Connect a 75-ohm cable between the A/B/C signal switch output and the VHF antenna input on the TV set.
9. Connect a coax cable with RCA phono plugs between the video output of VCR 1 and the video input of VCR 2. Use phono jack to miniplug adapters on either end as required.
10. Connect a coax cable with RCA phono plugs between the audio output jack of VCR 1 and the audio input jack of VCR 2. Use phono jack to miniplug adapters on either end as required.

To Record a Scrambled Channel While Viewing an Unscrambled Channel

1. Apply power to all units except VCR 2.
2. Tune VCR 1 to the converter-decoder output (channel 3 or 4).
3. Set the A/B/C switch to B.

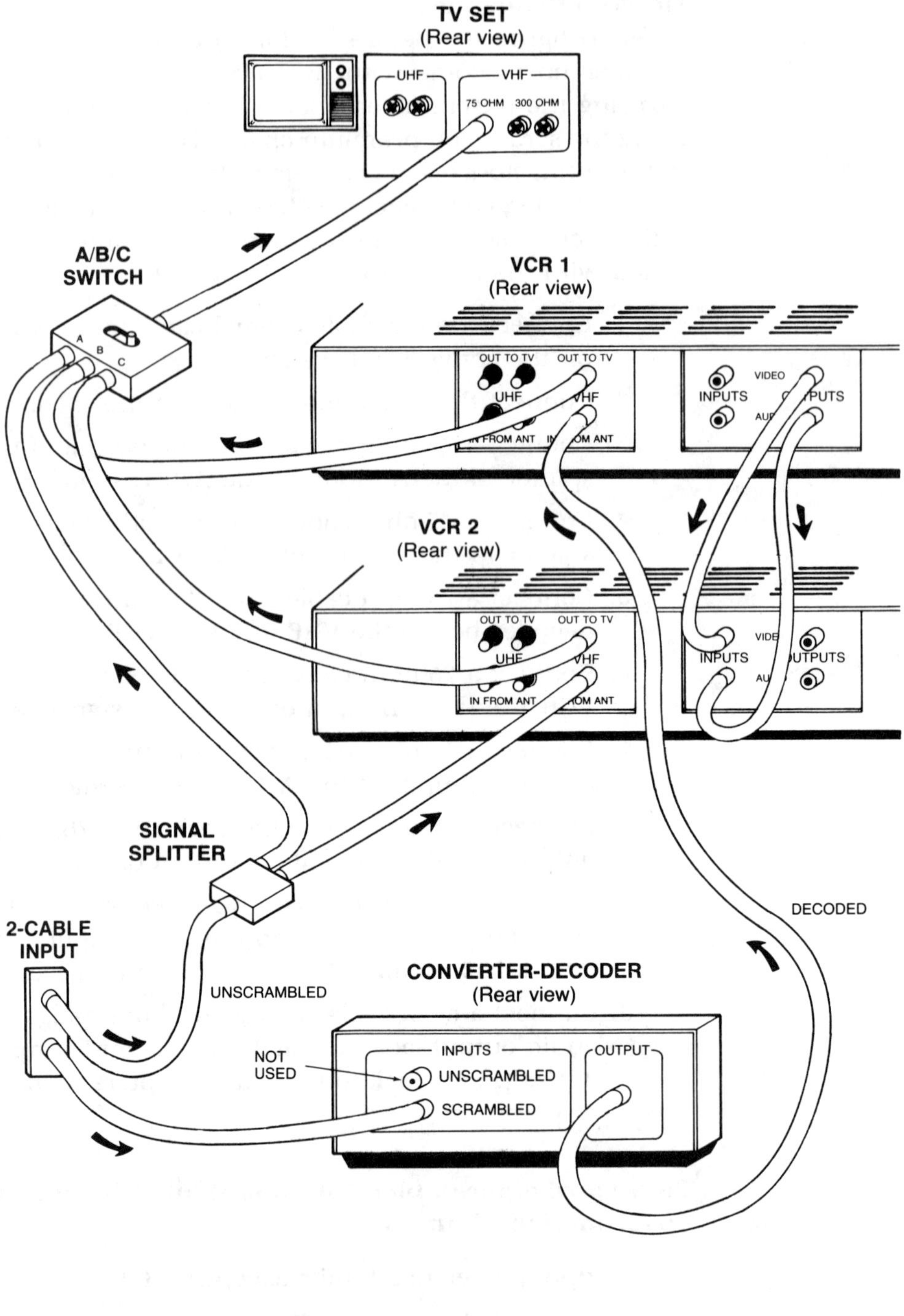

Figure 6-9
Hookup 19.

4. Tune the scrambled channel to be recorded with the converter-decoder.
5. Set the TV/VCR switch on VCR 1 to VCR.
6. Press the record button(s) on VCR 1.
7. View the TV to verify the channel to be recorded.
8. Set the A/B/C switch to A.
9. Tune the unscrambled channel to be viewed on the TV set.

To Record an Unscrambled Channel While Viewing a Scrambled Channel

1. Apply power to all units except VCR 1.
2. Set the A/B/C switch to C.
3. Tune the unscrambled channel to be recorded on VCR 2.
4. Set the TV/VCR switch on VCR 2 to VCR.
5. Tune the TV to the output of VCR 2 (channel 3 or 4).
6. Press the record button(s) on VCR 2.
7. View the TV to verify the channel to be recorded.
8. Set the A/B/C switch to B.
9. Tune the scrambled channel to be viewed on the converter-decoder box.
10. Tune the TV set to the output of the converter-decoder (channel 3 or 4).

To View and Record the Same Scrambled Channel

1. Apply power to all units except VCR 2.
2. Tune the channel to be viewed on the converter-decoder.
3. Set the A/B/C switch to B.
4. Tune the TV to the output of the converter-decoder.
5. Set the TV/VCR switch to TV.
6. Press the record button(s) on VCR 1.

To View and Record Any Unscrambled Channel

1. Apply power to all units except VCR 1.
2. Set the A/B/C switch to C.
3. Tune the channel to be viewed on VCR 2.

4. Tune the TV to the output of VCR 2 (channel 3 or 4).
5. Set the TV/VCR switch on VCR 2 to TV.
6. Press the record button(s) on VCR 2.

To Play a Prerecorded Tape While Recording a Scrambled Channel

1. Apply power to all units.
2. Insert the tape to be played in VCR 2. Put a blank tape in VCR 1.
3. Set the A/B/C switch to B.
4. Tune the scrambled channel to be recorded on the converter-decoder.
5. Set the TV to the output of VCR 1 (channel 3 or 4).
6. Press the record button(s) on VCR 1.
7. Set the TV/VCR switch on VCR 1 to VCR.
8. View the TV to verify the channel to be recorded.
9. Set the A/B/C switch to C.
10. Tune the TV to the output of VCR 2 (channel 3 or 4).
11. Press the play button on VCR 2.
12. View the prerecorded tape.

To View a Prerecorded Tape While Recording an Unscrambled Channel

1. Apply power to all units.
2. Put the tape to be viewed in VCR 1. Put a blank tape in VCR 2.
3. Tune VCR 2 to the unscrambled channel to be recorded.
4. Tune the TV to the output of VCR 2 (channel 3 or 4).
5. Set the TV/VCR switch on VCR 2 to VCR.
6. Set the A/B/C switch to C.
7. Press the record button(s) on VCR 2.
8. View the TV to verify the channel to be recorded.
9. Set the A/B/C switch to B.
10. Tune the TV to the output of VCR 1 (channel 3 or 4).
11. Press the play button on VCR 1.
12. View the prerecorded tape.

To Copy a Tape

1. Apply power to all units except the converter-decoder.
2. Put the tape to be recorded in VCR 2. Put a blank tape in VCR 1.
3. Tune the TV to the output of VCR 1.
4. Set the A/B/C switch to B.
5. Press the play button on VCR 2.
6. Press the record button(s) on VCR 1.

Important Notice

If during your hookup you must disconnect the incoming cable from the converter-decoder box, all or some of your cable channels may be disabled. Some cable TV systems sense the disconnection and turn off your cable service. After you have reconnected the cable, the signal is usually restored after a period of time. To have your service restored immediately, call your cable service company.

7 Special Hookups

The hookups covered in chapters 5 and 6 solve most interconnection problems. However, you may want to perform operations involving the interconnection of special equipment. This chapter outlines the more popular special interconnections, including the connection of your TV set to an existing stereo hi-fi system, copying (dubbing) tapes, and using a video camera.

Hookup 20—Connecting a TV to a Stereo System

Equipment
TV set with stereo capability and audio output jacks, stereo hi-fi system with amplifier and speakers.

Accessories
A pair of audio coax cables with RCA phono connectors on each end. Color-coded cables or connectors should be used. Place equipment in position and note cable lengths needed before purchasing cable.

Advantages
Can play TV stereo sound through an existing hi-fi system for improved quality and performance. Can use the headphone capability of stereo system.

Disadvantages

Cannot use the hi-fi system separately while using it for TV stereo. Cannot use the volume control on the TV remote control for setting the sound volume on the stereo amplifier.

Hookup Procedure

Refer to figure 7-1 before hooking up the equipment.

1. Connect one end of each audio cable to the stereo left and right audio outputs on the back of the TV set.
2. Connect the other ends of the audio cables to a free pair of auxiliary inputs on the hi-fi system amplifier. Be sure to connect left output to left input and right output to right input. Use the color code as a guide. *Note:* Most stereo amplifiers provide several pairs of auxiliary inputs for unspecified equipment. These are usually labeled AUX 1, AUX 2, and so on. The pairs of inputs are labeled Left and Right. To use the inputs, a front panel selector switch on the stereo amplifier must be placed in the correct position so that the

Figure 7-1
Hookup 20.

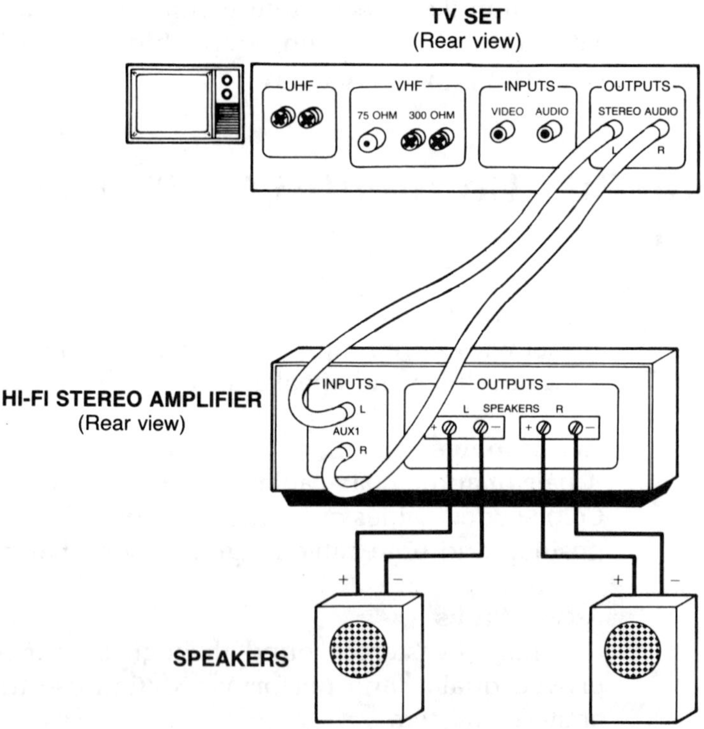

auxiliary inputs will be passed through the amplifiers and to the speakers.

To Use the Stereo System With the TV

1. Apply power to the TV set and stereo amplifier.
2. Set the switch on the back of the TV set, if used, to enable the stereo outputs and disable the internal TV set speakers.
3. Tune the TV set to the desired channel.
4. Set the input selector switch on the stereo amplifier to the auxiliary inputs used.
5. Adjust the volume and balance with stereo amplifier controls.

Hookup 21—Connecting a Stereo Decoder

This hookup covers two versions. Version A uses a VHF/UHF antenna input. Version B uses a cable TV input.

Equipment
TV set, VCR, stereo decoder, hi-fi stereo amplifier with speakers.

Accessories (Version A)
VHF/UHF antenna, signal splitter, two UHF/VHF band separators, three 75-ohm coax cables, one 300-ohm twin lead cable, two audio cables with RCA phono connectors. Place equipment in position and note cable lengths needed before purchasing cable.

Accessories (Version B)
Cable input, signal splitter, three 75-ohm coax cables, two audio cables with RCA phono connectors. Place equipment in position and note cable lengths needed before purchasing cable.

Advantages
Can hear stereo sound from over-the-air or cable signals. Can view one channel while recording another (version A).

Disadvantages
Cannot record stereo sound on the VCR. Cannot record one channel while viewing another (version B).

Hookup Procedure (Version A)

Refer to figure 7-2 before hooking up equipment.

1. Connect coax cable from the VHF/UHF antenna to a signal splitter.
2. Connect both outputs of the signal splitter to band separators.
3. Connect the VHF (coax) and UHF (twin lead) outputs from one band separator to the RF antenna inputs of the VCR.

Figure 7-2
Hookup 21, version A.

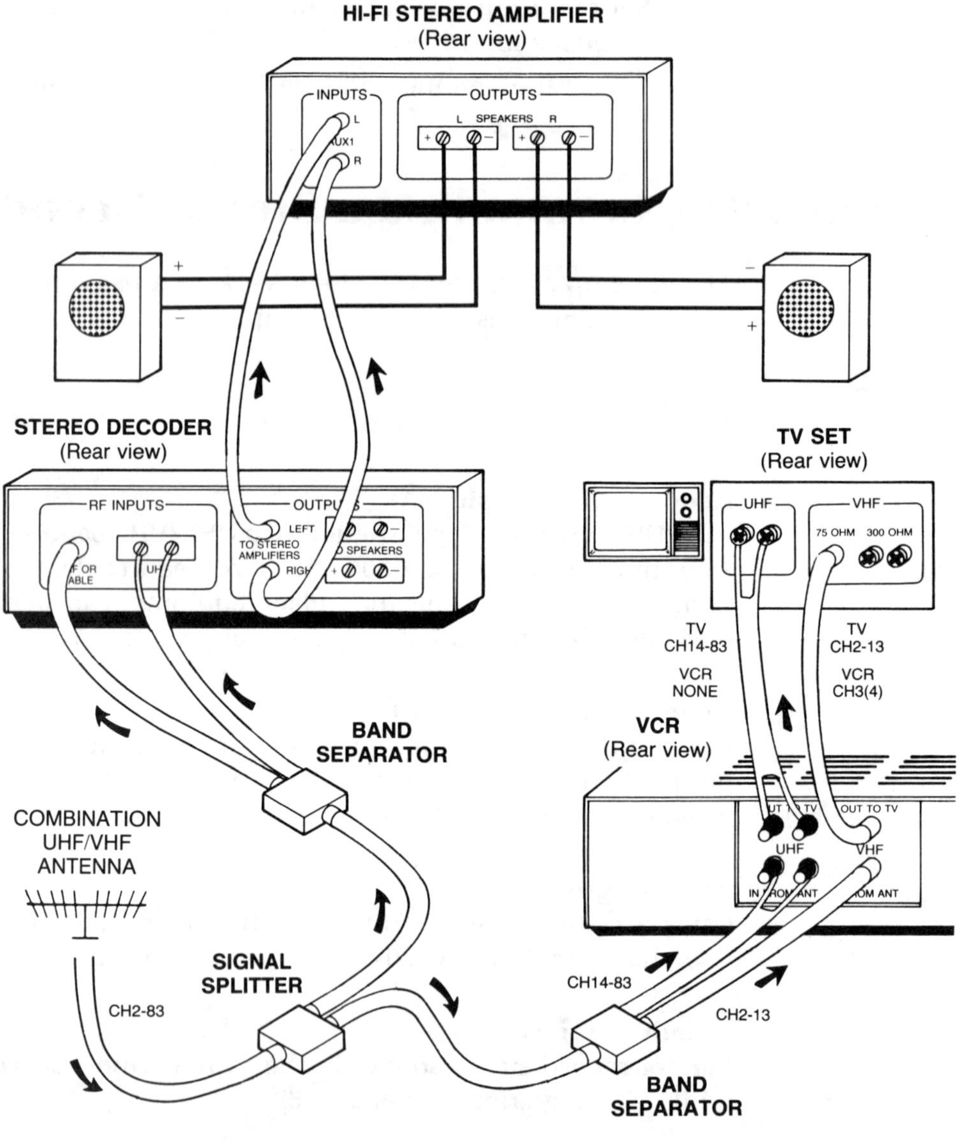

4. Connect the VHF and UHF outputs from the other band separator to the RF antenna inputs of the stereo decoder.
5. Connect the UHF output of the VCR to the UHF input on the TV set with a twin lead cable.
6. Connect the VHF output of the VCR to the VHF input on the TV set with a coax cable.
7. Connect the left and right outputs of the stereo decoder to the left and right connectors of one of the auxiliary inputs on the hi-fi stereo amplifier.

Hookup Procedure (Version B)

Refer to figure 7-3 before hooking up equipment.

1. Connect the cable input to the input connector on the cable converter-decoder box.
2. Connect the output of the cable converter-decoder box to the input of a signal splitter with a coax cable.
3. Connect one of the signal splitter outputs to the VHF input on the VCR with a coax cable.
4. Connect the VHF output of the VCR to the VHF input on the TV set with a 75-ohm cable.
5. Connect the other output of the signal splitter to the VHF input on the stereo decoder with a coax cable.
6. Connect the left and right outputs of the stereo decoder to the left and right phono connectors on one of the auxiliary inputs on the hi-fi stereo amplifier. *Note:* Some stereo decoders have built-in audio amplifiers and can be connected directly to speakers. As an alternative to using a separate hi-fi stereo system, connect the left and right speaker outputs of the stereo decoder directly to the speakers.

To Listen to Stereo Sound (Version A or B)

1. Turn on power to the TV set, stereo decoder, and hi-fi stereo amplifier. Power to the VCR should be off.
2. Select the channel to be viewed on the TV set and stereo decoder. Both should be set to the same channel.
3. Turn the TV sound volume all the way down.
4. Set the stereo amplifier input switch to the auxiliary inputs to which the stereo decoder is connected.

Figure 7-3 Hookup 21, version B.

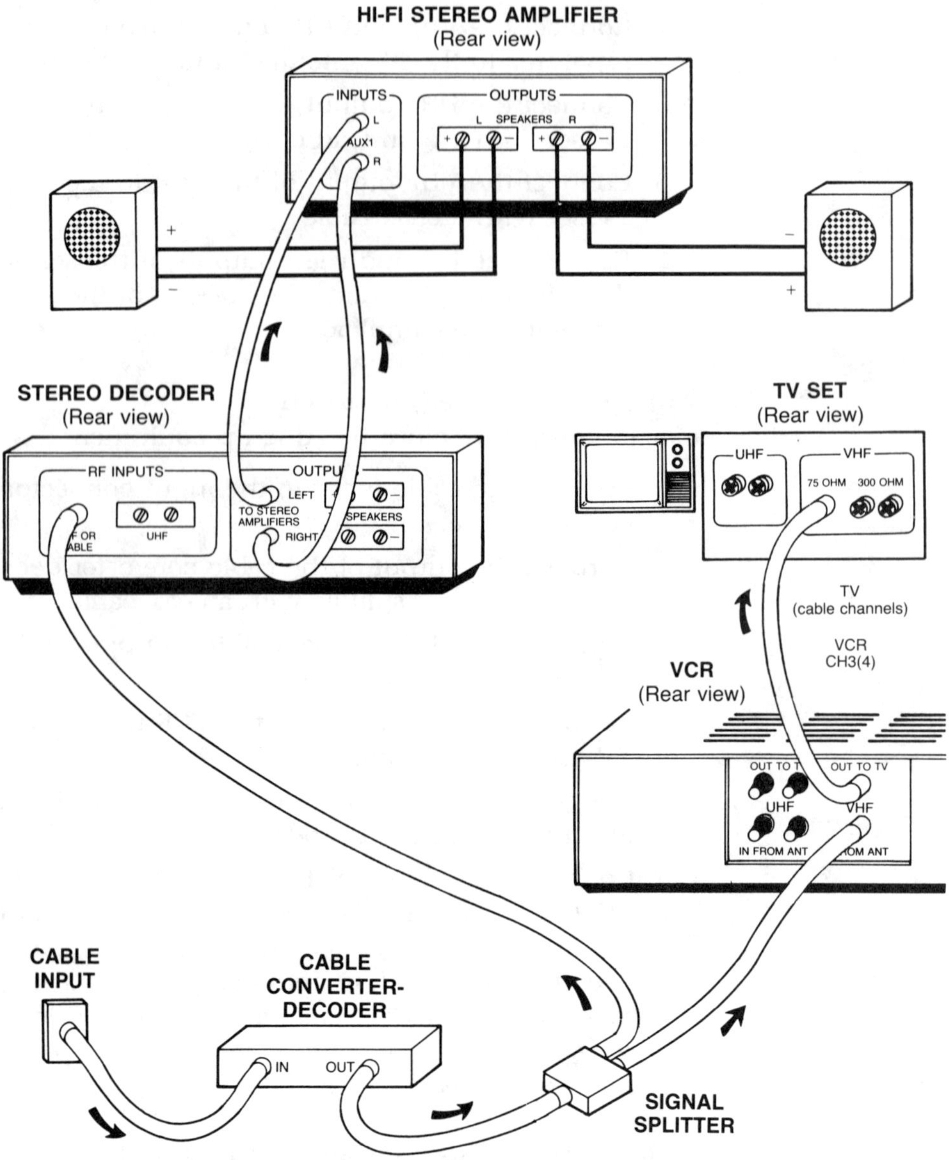

5. Adjust the volume on the hi-fi stereo amplifier to the desired listening level.

To View One Channel With Stereo Sound While Recording Another Channel (Version A Only)

1. Apply power to the TV set, VCR, stereo decoder, and hi-fi stereo amplifier system.

2. Select the channel to be viewed with the TV set tuner.
3. Set the stereo decoder to the same channel as the TV set.
4. Set the VCR channel to the station to be recorded.
5. Turn the TV sound volume all the way down.
6. Set the stereo amplifier input switch to the auxiliary channel to which the stereo decoder is connected.
7. Set the VCR TV/VCR switch to TV.
8. Adjust the volume control on the hi-fi stereo amplifier for the desired listening level.

To View and Record the Same Channel (Version B Only)

1. Turn on power to the TV set, VCR, cable converter-decoder, stereo decoder, and hi-fi stereo amplifier.
2. Select the desired channel with the cable converter-decoder box.
3. Set the VCR and TV to channel 3 or channel 4.
4. Set the stereo decoder to channel 3 or channel 4.
5. Turn the TV set volume all the way down.
6. Set the stereo amplifier input selector switch to the auxiliary inputs to which the stereo decoder is connected.
7. Adjust the volume of the stereo amplifier for the desired listening level.

Hookup 22—Copying a Video Tape

Three versions of this hookup can be implemented. Version A uses the video and audio outputs and inputs of the VCRs. Version B uses the VCR VHF outputs and inputs. Version A, which uses the separate video and audio cables, is preferred because all modulation and demodulation processes are bypassed, thereby improving the quality of the copy. Version C uses a video processor between the VCRs to enhance the video image.

Equipment
Two VCRs (any combination of VHS or Beta format), video processor (version C only).

Accessories (Version A)
Two coax cables with RCA phono connectors, one coax cable with F-plugs, two or four phono jack to miniplug adapters if Beta VCRs are used. Place equipment in position and note cable lengths needed before purchasing cable.

Accessories (Version B)
One 75-ohm coax cable with F-plugs. Place equipment in position and note cable lengths needed before purchasing cable.

Accessories (Version C)
Four coax cables with RCA phono connectors, two or four phono jack to miniplug adapters if Beta VCRs are used, one coax cable with F-plugs. Place equipment in position and note cable lengths needed before purchasing cable.

Advantages
Can copy a video tape from one VCR to another. Can accommodate any combination of VCRS: VHS-to-VHS, VHS-to-Beta, Beta-to-Beta, or Beta-to-VHS. Can view tape being copied. Can enhance the quality of the video using a video processor (version C).

Disadvantages
Although version B requires a simpler hookup, the use of modulation and demodulation can degrade picture and sound quality during copying. Cannot perform any other video operations while copying the tape.

Hookup Procedure (Version A)
Refer to figure 7-4 before hooking up equipment.

1. Connect one end of a 75-ohm cable to the video output jack of the source VCR. Connect the other end of the video cable to the video input jack on the recording VCR. Use phono jack to miniplug adapters if required.
2. Connect a 75-ohm cable from the audio output of the source VCR. Connect the other end of the cable from the source VCR audio output to the recording VCR audio input jack. Use phono jack to miniplug adapters if necessary.
3. Connect the VHF output jack on the recording VCR to the VHF antenna input on the TV with a coax cable with F-plugs.

Figure 7-4
Hookup 22,
version A.

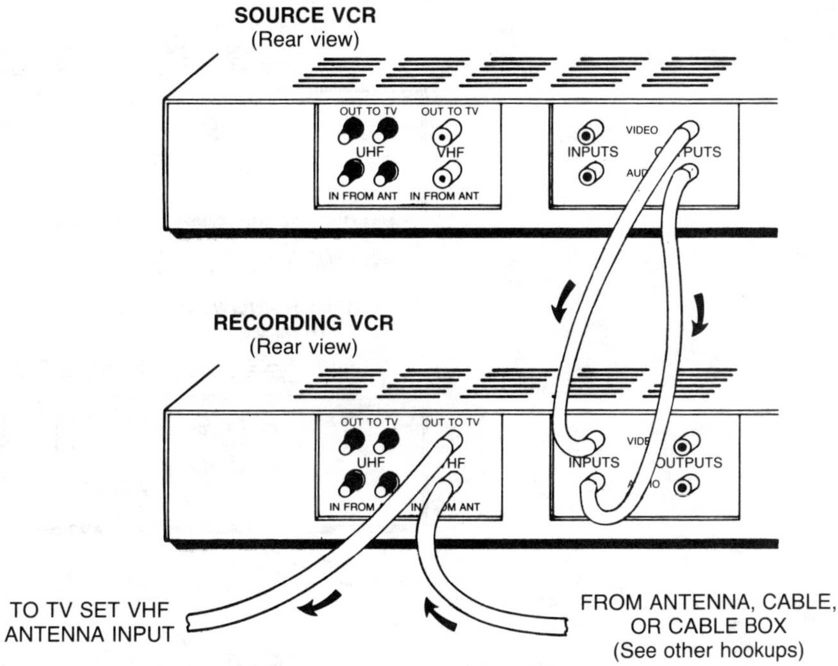

4. The VHF input jack of the recording VCR can be connected to an antenna, a cable, or a cable converter box if the VCR is part of another hookup.

Hookup Procedure (Version B)

Refer to figure 7-5 before hooking up equipment.

1. Connect one end of a 75-ohm cable with an F-plug to the VHF output jack on the source VCR.
2. Connect the other end of the 75-ohm cable with an F-plug to the VHF input jack on the recording VCR.

Hookup Procedure (Version C)

Refer to figure 7-6 before hooking up equipment.

1. Connect a 75-ohm cable to the video output jack of the source VCR. Use a phono jack to miniplug adapter if necessary. Connect the other end of the video cable to the video input of the video processor.
2. Connect a 75-ohm cable to the audio output jack of the source VCR. Use a phono jack to miniplug adapter if

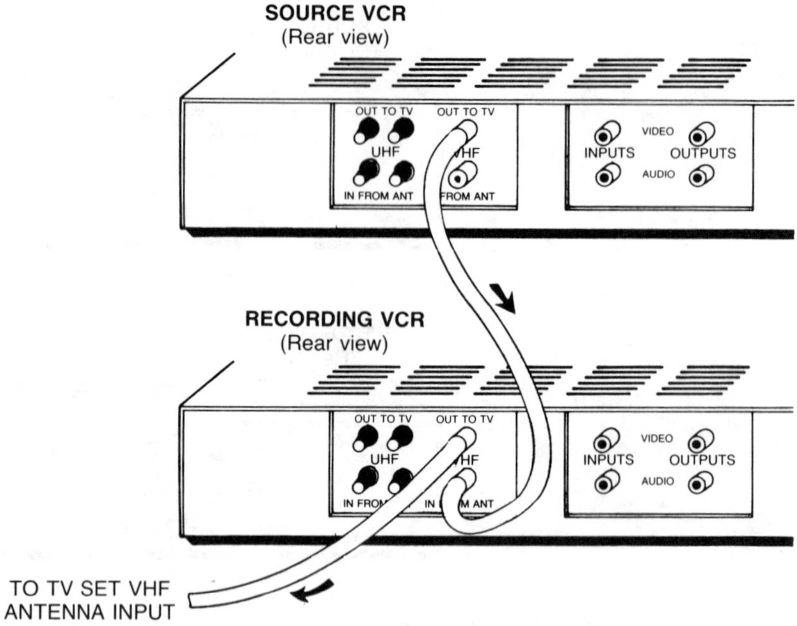

Figure 7-5 Hookup 22, version B.

necessary. Connect the other end of the audio cable to the audio input jack on the video processor.

3. Connect a 75-ohm cable to the video output of the video processor. Connect the other end of the video cable from the processor to the video input jack on the recording VCR. Use a phono jack to miniplug adapter if necessary.
4. Connect one end of a 75-ohm cable to the audio output jack on the video processor. Connect the other end of the audio cable from the processor to the audio input jack on the recording VCR. Use a phono jack to miniplug adapter if necessary. *Note:* If the VCRs are stereo units, dual (left and right) audio cables are needed between the source VCR and video processor and the video processor and recording VCR. Most processors can handle stereo audio.
5. Connect a coax cable with F-plugs between the VHF antenna output on the recording VCR and the VHF antenna input on the viewing TV set.

To Copy a Video Tape (Versions A and B)

1. Apply power to both VCRs and the TV set monitor.
2. Insert the tape to be copied in the source VCR. Be sure the tape is rewound.

Figure 7-6
Hookup 22,
version C.

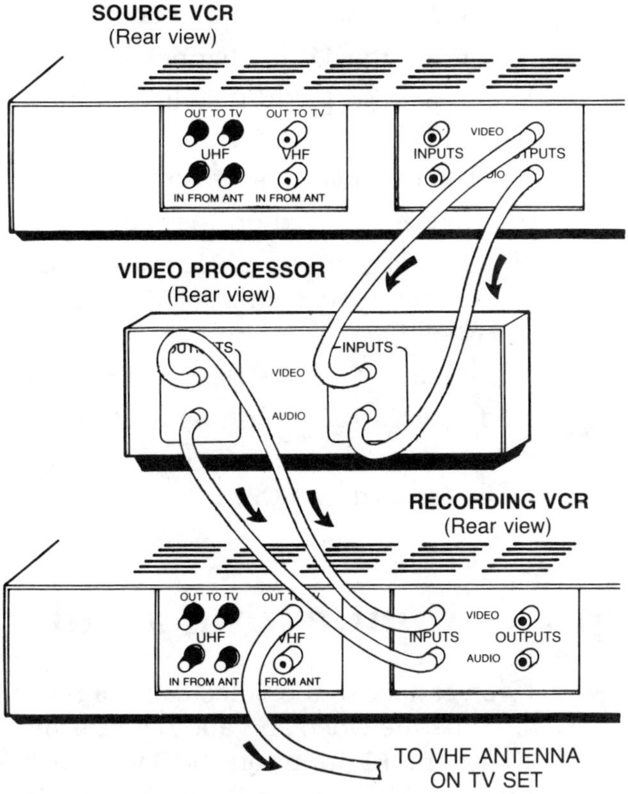

3. Insert a blank video tape in the recording VCR.
4. Set the TV set to channel 3 or channel 4.
5. Set the TV/VCR switch on the recording VCR to TV.
6. Press the play button on the source VCR.
7. Press the record button on the recording VCR.
8. When copying is complete, press the off buttons on both source and recording VCRs.
9. Rewind both tapes by pressing rewind on both VCRs.
10. Verify that the tape was copied by pressing play on the recording VCR.

To Copy a Tape While Providing Image Enhancement With a Video Processor (Version C)

1. Apply power to both VCRs, the TV set, and the video processor.
2. Insert the tape to be copied into the source VCR.

3. Insert a blank video tape into the recording VCR.
4. Set the TV to channel 3 or 4.
5. Set the TV/VCR switch to the TV position on the recording VCR.
6. Press play on the source VCR.
7. Press record on the recording VCR.
8. Adjust the controls on the video processor while observing the picture on the TV set. Set controls for best video enhancement.
9. After the tape has been copied, press stop on both VCRs.
10. Press rewind on both VCRs.
11. To verify that the tape has been copied, press play on the recording VCR.

Hookup 23—Connecting a Video Camera

Two versions of the video camera hookup are offered. Version A uses the separate video and audio inputs on the TV set if available. Version B uses the RF input on the TV set if the video and audio inputs are not available. Version B requires an RF modulator.

Equipment (Version A)
Video camera, TV set with video and audio inputs.

Equipment (Version B)
Video camera, TV set, RF modulator.

Accessories (Version A)
Two 75-ohm cables with RCA phono connectors, any special adapters as required by the video camera. Place equipment in position and note cable lengths needed before purchasing cables.

Accessories (Version B)
Two 75-ohm coax cables with RCA phono connectors, any adapters as required by the video camera, two 75-ohm coax cables with F-connectors, one A/B switch, matching transformer if required. Place equipment in position and note cable lengths needed before purchasing cables.

Advantage
Can monitor scenes with a video camera.

Disadvantage
Cannot view other signal sources while monitoring with a video camera.

Hookup Procedure (Version A)
Refer to figure 7-7 before hooking up equipment.

1. Connect one end of a coax cable to the video output jack of the camera. Connect the other end of the video cable to the video input jack on the TV set.
2. Connect one end of a 75-ohm cable to the audio output jack on the camera. Connect the other end of the audio cable to the audio input jack on the TV set.
3. Set the switch on the TV (if present) to external video input.

Hookup Procedure (Version B)
Refer to figure 7-8 before hooking up equipment.

1. Connect one end of a 75-ohm cable to the video output jack on the video camera. Connect the other end of the video cable to the video input on the RF modulator.
2. Connect one end of a 75-ohm cable to the audio output jack on the video camera. Connect the other end of the audio cable to the audio input jack on the RF modulator.

Figure 7-7
Hookup 22, version A.

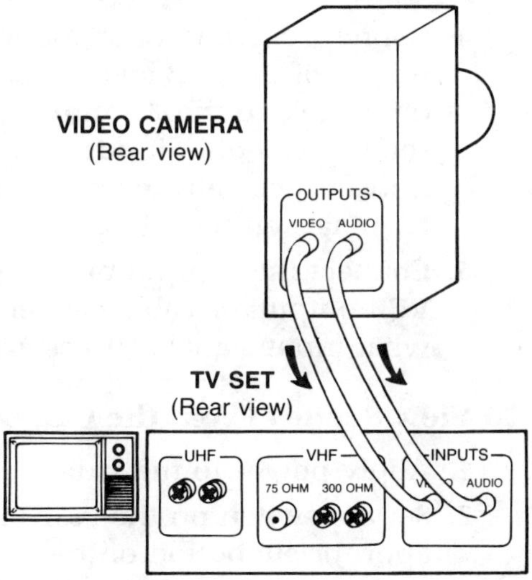

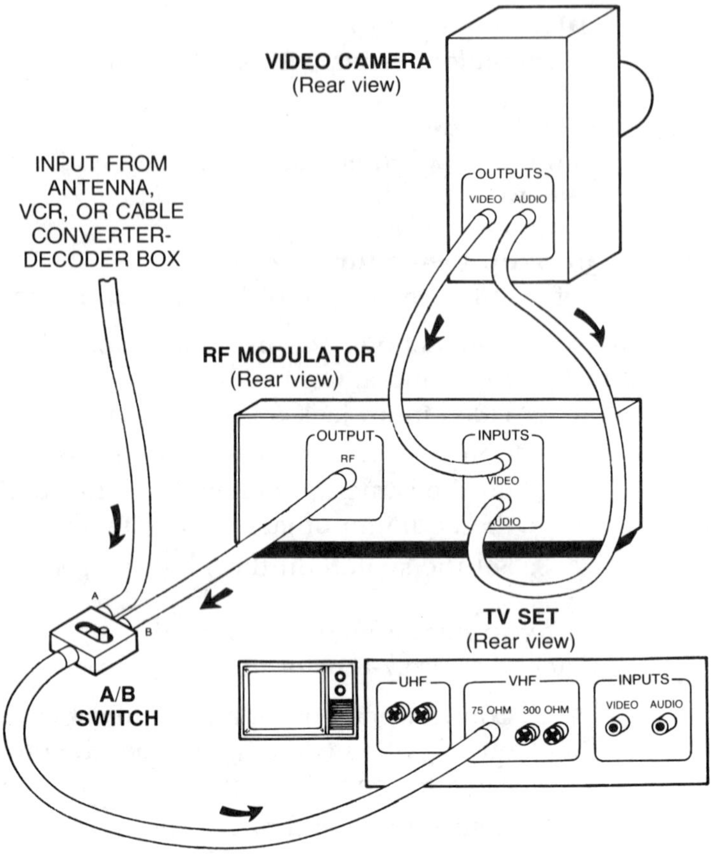

Figure 7-8 Hookup 22, version B.

3. Connect the RF output of the RF modulator to the B input of an A/B switch using a 75-ohm cable with F-connectors.
4. Connect one end of a 75-ohm cable with F-connectors to the output of the A/B switch. Connect the other end of the 75-ohm cable to the 75-ohm VHF input on the back of the TV set. If no 75-ohm input is available, use a matching transformer between the cable and the 300-ohm input terminals on the TV set.
5. Connect the A input of the A/B switch to the VHF antenna, VCR output, or cable converter box output depending upon what input signal sources are available.

To View Scenes From the Camera (Version A)

1. Apply power to the video camera and TV set.
2. Set the switch on the rear of the TV set or depress the appropriate button on the remote control to enable the

video/audio inputs on the TV set and disable the internal tuner. Check your TV set instruction manual for details.

3. Set the controls on the camera as required. See the manual.

To View Scenes From the Camera (Version B)

1. Apply power to the video camera, TV set, and RF modulator.
2. Set the TV to channel 3 or 4 as required by the RF modulator.
3. Set the A/B switch to the B position.
4. Adjust the controls on the camera as required. See the manual.

Hookup 24—Connecting a Video Game

Equipment
TV set, VCR, video game.

Accessories
A/B signal switch, three 75-ohm coax cables with F-plugs, 75-ohm-to-300-ohm matching transformer (if needed), RF modulator (if needed). *Note:* Most video games have a built-in RF modulator. The output can be 300 ohms or 75 ohms. Use a matching transformer between the video game RF output and the A input to the switch if necessary.

Advantage
Can record a channel while playing a video game.

Disadvantage
Cannot view any other signal while playing a video game.

Hookup Procedure
Refer to figure 7-9 before hooking up equipment.

1. Connect the coax cable from the antenna or cable converter box output to the VCR VHF ANT input jack.
2. Connect one end of a 75-ohm cable to the VHF output jack on the VCR. Connect the other end to the B input on the signal switch.

Figure 7-9
Hookup 24.

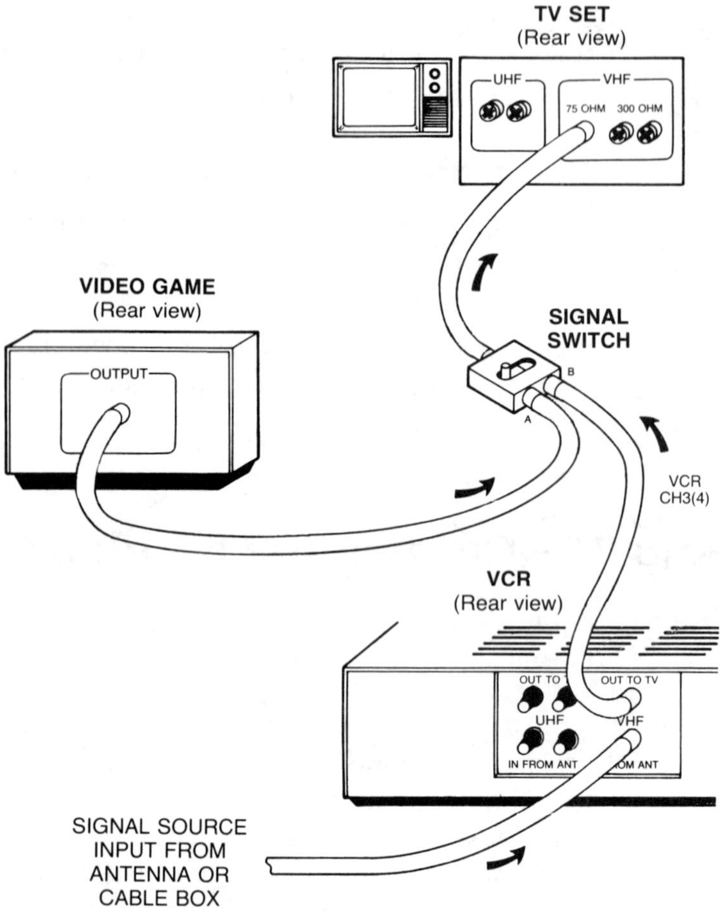

3. Connect one end of a 75-ohm cable to the video game output jack. Connect the other end to the A input of the signal switch.
4. Connect the output jack of the A/B switch to one end of a 75-ohm cable. Connect the other end of the cable to the VHF antenna input jack on the TV.

To Play a Video Game

1. Turn on the TV and video game power. Turn off the VCR power.
2. Set the signal switch to A.

To View a TV Program Without Recording

1. Turn on the TV power. Turn off the video game and VCR power.
2. Select the channel to view with the TV if the cable box is *not* used.
3. Set the TV to channel 3 or 4 and select the channel to view with the cable box, if used.

To View a Prerecorded Tape

1. Be sure the VCR and TV power switches are on. Turn off the video game.
2. Set the signal switch to B.
3. Set the TV/VCR switch on the VCR to VCR.
4. Tune the TV to the output channel of the VCR (channel 3 or 4).
5. Press the play button on the VCR.

To Record and View the Same Channel

1. Be sure the TV and VCR power switches are on. Turn off the video game.
2. Set the signal switch to B.
3. Tune the VCR to the down converter output channel.
4. Set the TV/VCR switch on the VCR to VCR.
5. Tune the TV to the output channel of the VCR.
6. View the TV to verify the channel to be recorded.
7. Press the record button(s) on the VCR.

To View One Channel While Recording Another Channel

1. Be sure the TV and VCR power switches are on.
2. Set the signal switch to B.
3. Tune the VCR to the channel to record.
4. Set the TV/VCR switch on the VCR to VCR.
5. Tune the TV to the output channel of the VCR (channel 3 or 4).
6. View the TV to verify the channel to record.
7. Press the record button(s) on the VCR.
8. Set the TV/VCR switch on the VCR to TV.
9. Tune the TV to the desired channel to view.

To Play a Video Game While Recording a Channel

1. Turn on all power switches.
2. Set the signal switch to B.
3. Set the TV set to channel 3 or 4.
4. Select the channel to record with the VCR.
5. Press the record button(s) on the VCR.
6. View the TV to verify the output channel to record.
7. Set the signal switch to A.
8. Play the video game.

Creating Your Own Hookups

As you might suspect, there is a huge number of possible hookups. It's impossible to cover all of them here, but we have included the most popular and useful hookups. In most cases, you should find one that will fit your situation.

But there are always exceptions. You may have an unusual selection of equipment. You may want to perform some functions not possible with the hookups given. In these cases, you need a special hookup. Chances are you can create it yourself. Here's how:

1. List all your equipment.
2. List the operations you want to perform.
3. Find the hookup or hookups in this book that are closest to your desired configuration and operation. Use the Contents and Table 4-1 as guides in identifying these hookups.
4. Read the relevant hookups.
5. Modify the closest hookup to include your special equipment or operations. To do this:
    a. Combine smaller parts of other hookups relative to your equipment.
    b. Add new accessories as required. Refer to chapters 2 and 3 for details on equipment and accessories. After you know how they work, you should be able to reason out the connections.
    c. Refer to your equipment manuals for help or hints.
6. Experiment. Try out your hookup. Play around with different cable connections, switch positions, control settings, etc.

8 Maintenance and Troubleshooting

In general, there is little you can do to repair the equipment in your video system. Most units are generally trouble-free and should give exceptional performance for a long time. But occasionally you will have troubles. Major equipment breakdowns should be fixed by qualified service personnel. TV sets, VCRs, and other video equipment are complex. Do not attempt to service them yourself.

But you can correct some problems in the system. In this brief chapter, we outline some maintenance guides and troubleshooting procedures you can follow to locate difficulties.

Maintenance

Maintenance refers to actions you can take to ensure a fully operational system. In this case, the only maintenance you can perform is to keep the system clean and ensure that all connections are tight.

To clean your video system, dust it frequently. Dust build-up can cause damage to delicate mechanical units such as VCRs. Dust can also block heat openings and cause heat to build up, which, in turn, can produce component failures. It is a good idea to wipe down all your equipment, using a lint-free cloth, at least once a month. Also, do *not* use furniture polish.

Another thing you can do to ensure a reliable system is to check all connections periodically. Verify that all F-connectors are screwed on tightly and that other connections are secure. Loose screw terminals or corrosion can cause a poor connection. Poor connections generally cause noise in the signal or no operation at all. Once a month, check all connections by tightening screw terminals and rotating press-on or slip-on F-connectors and phono connectors to ensure good electrical contact. If you move your equipment frequently, broken and loose connections can occur.

It is also a good idea to clean all connectors once or twice a year. Exposure to air causes all contacts and connectors, even gold ones, to oxidize. This increases electrical resistance and can introduce signal losses and noise. It can reduce picture quality or cause noise and hum in the audio.

To clean the connectors, first unplug all units. Then disconnect all cables and connectors, one end at a time, and clean them. Don't pull all cables at once unless you label them because you will have a major reconnection project later.

To clean the connectors, dampen a cloth with isopropyl alcohol (*not* rubbing alcohol) or a metal contact cleaner such as Cramolin, Tweek, or other brands found in electronic stores. Clean all metal surfaces, including AC plug prongs. Use a cotton swab to get into tight places. Then replace all connectors.

Troubleshooting

In general, good maintenance will keep your system operating without difficulty. But occasionally some problem occurs. In such cases, you will need to perform some troubleshooting to track down the problem.

Most video system problems are in one of three categories: cockpit problems, hookup defects, and dirty VCR heads. Let's take a look at each in more detail.

Cockpit Problems

A cockpit problem is basically an operator problem. It means you are using the system incorrectly or you are attempting to do something that the system is incapable of doing.

The most common cockpit error is an incorrect control setting. Some video systems described in this book are rather sophisticated. The newer TV sets, VCRs, and other units have multiple controls that must be set correctly.

The first thing to do when you have a problem is to check all control settings. If necessary, go back to your TV, VCR, or other instruction manual to be sure you have the correct settings. For example, one of the most common problems is to have the VCR's TV/VCR switch in the wrong position. If you are using A/B switches in your hookups, check to see that each switch is in the correct position for the operation desired. Also check that your TV set, VCR, RF modulator, or other unit is set to channel 3 or 4 as required.

A cockpit problem is in a sense no problem at all. It means all equipment is operating and the hookup is correct. Your job is to learn how to set the various controls properly for the desired operation. Becoming more familiar with the system usually solves this problem.

Interconnection Problems

One of the most common causes for system failure is an interconnection defect. Over time, connections can become dirty or wires can be broken. Dirty connections usually manifest themselves as a noisy or snowy picture or scratchy sound or hum in the audio. A poor interconnection introduces attenuation and signal distortion. You may notice horizontal streaks across the picture or a weak signal with lots of snow.

Sometimes F-connectors and RCA phono connectors become dirty. They oxidize or corrode if they have been in place for some time. This usually introduces noise and attenuation. Simply disconnect the connector and reconnect it. By moving the plug in and out of the jack, you remove the oxidation or corrosion. Then be sure the connection is tight.

If the connections are badly oxidized, clean them using the procedures described earlier. If they are badly corroded, replace them.

Sometimes the connection is not dirty, but simply loose. A screw terminal could be loose and a spade lug is not making connection. Simply tightening the screw usually solves the problem. Moving equipment can loosen or partially disconnect a connector. If the system is not performing correctly, check that all cables are in their proper places.

A very common problem is a broken 300-ohm twin lead wire. The wire conductors in twin lead are thin and very delicate. If moved around too much, they can break easily. This problem can be repaired quickly by simply stripping the cable again, exposing the conductor, and reconnecting it. Every time you move or clean the equipment, check all the interconnections to be sure they are firm and tight.

Dirty VCR Heads

A common problem in VCRs is when the record/playback heads that ride against the tape become dirty. Oxidation can form or the magnetic oxide on the tape can rub off and build up on the head. The result is usually poor picture quality on playback. This shows up as severe distortion or tearing. Less severe troubles include horizontal streaks and noise. These problems can usually be cleared up by cleaning the VCR heads.

Many VCR manufacturers recommend that you do not clean the heads yourself. Sometimes, this is good advice. But in most cases, you can solve the dirty head problem yourself with a cleaning tape.

The most thorough way to clean the heads is to open the VCR unit and clean them using the special chemicals supplied in a VCR cleaning kit. Such kits are available at electronic and video stores. They usually contain a special brush, mirror probe, and a cleaning chemical, such as Freon TF made by DuPont. The main problem with this approach is that it requires you to open the VCR unit. Often, this will void the warranty on your VCR.

The fastest and easiest way to clean your VCR heads is to purchase a cleaning tape. You simply load the cleaning tape into the VCR as you would an ordinary tape, then press the play button. The nonabrasive tape holds a cleaning chemical that is passed through the system and over the heads. Cleaning usually only takes ten to thirty seconds. Before using some brands of cleaning tapes, a special cleaning fluid must be applied. Again, buy one of the commercial tapes at a video or electronic dealer and follow the directions.

In general, you could clean the heads every now and then as a maintenance procedure. But we don't recommend this. Follow the old advice, "If it's not broken, don't fix it." Only clean the VCR heads when you notice excessive noise or distortion and loss of color quality.

Glossary

75-Ohm Lead—Coax often used for TV and video equipment connection.

300-Ohm Lead—The type of signal lead often used for UHF signal connections. Also see *twin lead.*

Adapter—A device such as a special connector that allows accessories with different input or output connections to be used together.

Adapter Jack—A connector designed to join mismatched connectors and cables.

Add-On Accessory—An additional component of a video system.

Amplifier—A device that increases signal strength to make up for the losses of multiple hookups.

Antenna—A wire or rigid metal conductor that collects or radiates radio frequency energy.

Attenuation—A reduction of signal strength. Loss of signal level.

Audio—A sound signal; electrical voice or music.

Audio-Video Signal—Picture and sound converted into electricity that must be connected by wires to devices for recording or reconstructing the picture and sounds.

Balun—See *impedance-matching transformer.*

Band—Any range of frequencies that lies between two defined limits. A group of channels assigned by the FCC to a particular type of TV service.

Band Combiner—A device that combines ranges of frequencies from different sources, such as the UHF and VHF TV channel onto one output cable.

Band Separator—A frequency selective device that divides UHF, VHF, and FM from each other; same as band splitter.

Band Splitter—A frequency selective device that divides UHF, VHF, and FM from each other; same as band separator.

Band Splitter-Matching Coupler— A device that functions as both a band splitter and a matching transformer.

Band Switch—A switch used to select any one of the frequency bands in which an electronic apparatus can operate.

Block Converter—A device that shifts a range of channels either up or down to new channels.

Cable—An assembly of two or more wire conductors usually within a protective sheath and arranged so that the conductors can be used separately or in groups.

Cable Box—See *cable converter.*

Cable Converter—A device supplied by a cable company for tuning cable channels and changing the selected channel to a channel a regular TV set can receive.

Cable Programming—Programs available from a cable company that often require a special tuning box for the cable channels, as well as decoding for extra-cost channels.

Cable Ready—A TV set or VCR with a built-in tuner capable of receiving over-the-air signals from an antenna or cable TV signals without a cable converter box.

Cable TV—A television system that transmits standard programs and scrambled TV signals directly by cable to paid subscribers.

Camcorder—A combination video camera and recorder. A portable video camera/recorder unit.

Carrier—A high-frequency signal upon which video or audio signals are impressed by modulation to transmit them by radio waves from one place to another.

CATV—Abbreviation for cable TV.

Channel—A standard range of radio frequencies for transmission and reception of TV signals; it is not the station carried by the channel.

Coaxial Cable (Coax)—A transmission line with a center wire and a concentric outside conductor shield.

Combiner—See *band combiner.*

Conductor—A wire or metal element that carries electrical current with minimum resistance to current flow.

Connecting Cables—Twin lead, coax, or other wires or cables used to interconnect video equipment.

Connector—A metallic fitting on the end of a cable or on a piece of equipment used for facilitating interconnections. A plug or jack that joins cables with electronic units and other cables.

Converter Box—See *cable converter.*

Converter-Decoder—A cable converter that also restores scrambled channels.

Converter-Selector—A converter that also allows the selected channel to be converted to a channel a regular TV set can receive.

Coupler—A hookup accessory that allows two similar devices to operate from a single input. Also called a signal splitter.

CRT—Cathode ray tube. The picture tube or video display in a TV set or monitor.

Decoder—A device that restores a coded signal to an uncoded condition.

DIN Connector—A multipin (3 to 8 pins) connector that originated in Germany. Used for audio and video connections.

Direct Connection—An attachment without special extra tuners, decoders, or converters.

Distribution Switcher—A device that is a combination of RF amps, modulators-combiners, and remote control repeaters-extenders. See *video switcher.*

Down Converter—A device that changes a range of frequencies to a lower frequency range.

Dubbing—Copying a tape. Re-recording the video/audio on another tape.

Extra-Cost Channel—A premium channel that is scrambled or delivered only to subscribers.

F-Connector—A plug or jack commonly used to interconnect 75-ohm coax cable.

Frequency—The number of electromagnetic cycles or alternations completed in one second.

Ghosts—Unwanted second images next to, and possibly overlapping, the desired images. Usually caused by RF signal reflections from buildings, mountains, and so on.

Ground—A common metallic connecting point of zero voltage. Earth ground.

Hookup—The method or pattern of connecting the various components of a video system.

Hookup Accessory—A device that is used in hooking up video system equipment but is not a power source.

Impedance—The total opposition (resistance and reactance) a circuit offers to the flow of AC at a given frequency. The ratio of the potential difference across a circuit or element of a circuit to the current through the circuit or element.

Impedance-Matching Transformer—A transformer that matches the impedance of one device to the impedance of another.

Infrared—Electromagnetic waves with frequencies between visible light and radio waves. Used in wireless remote control and other transmission applications.

Input—The terminal jacks or receptacles provided for the introduction of an electrical signal into a system component.

Insertion Loss—Attenuation introduced by a matching transformer, band separator, splitter, filter, or other accessory.

Jack—A type of connector. The receptacle that accepts a plug.

Lead—A wire or cable between the components in a video system.

Loss—Attenuation of signal strength.

Matching Transformer—See impedance-matching transformer.

Miniconnector—Miniature plug or jack used in making audio connections in some equipment and video signal connections in Beta format VCRs.

Modulator—A device for combining signals to produce channels that can be broadcast and processed to recover the original signals. A circuit that causes the audio or video to modify an RF carrier signal for transmission.

Monitor—A device that converts a video signal into a picture on a TV screen. (Monitors are often also able to produce sound from an audio signal.) A TV set acts as a monitor to a VCR.

Mono (Monophonic)—Pertaining to audio information on one channel (as opposed to binaural or stereophonic) and usually referring to a one-speaker system.

Monochrome—Pertaining to or consisting of one color; for example, a black-and-white TV.

MTS—Multichannel TV sound. Stereo audio on broadcast TV.

Multiset Coupler—Same as a splitter.

On-Off Control—A switch that turns a device fully on or fully off. No other operating positions are available.

Output—The terminals, jacks, or receptacles provided for the exit of an electrical signal from a system component.

Output Channels—The selected bands of frequencies that a device produces.

Pay TV—A subscription service that provides a decoder for scrambled broadcast decoding of special programming.

Pay TV Decoder—A device for restoring a scrambled channel.

Phono Cable—A cable with plugs to match phono jacks.

Phono Connectors—Special plugs and jacks originally designed by RCA for connecting turntables and other audio equipment. Widely used in hi-fi stereo systems and in some video applications.

Plug—A connector that fits into a jack.

Premium Cable Channel—A cable channel that is usually scrambled, such as HBO, Cinemax, Disney, and Showtime. See *extra-cost channel.*

Radio Frequency (RF) Signal—The frequency that encompasses radio signal frequencies above 150 kHz and extending to infrared range.

RCA Connector—See *phono connectors.*

Receiver—A device such as a TV that selects and reproduces signals from a transmitter and converts electric waves into visible or audible form.

Remote Control—Any system of control performed from a distance and conveyed by intervening wires, sound (ultrasonics), light, infrared radiation, or radio.

RF—See *radio frequency signal.*

RF Amplifier—A device that increases the strength of RF signals.

SAP—Separate audio program. Special audio signal sometimes transmitted along with MTS sound.

Satellite Receiver—A device for playing rebroadcast TV from a satellite.

Scrambled Channel—A signal that is encoded so it is unusable unless a special decoder unit is used to restore it.

Selector—A device that tunes one program out of many. A filter.

Signal Combiner—A device for adding channels, bands, or frequencies. Combines multiple inputs into a single output.

Signal Splitter—A device that equally divides an input signal into two or more output signals for connection to other devices.

Signal Switch—A device for connecting electrical devices together in one of more than one possible ways. Used to select two or more signal sources or destinations.

Snow—Speckled interference in a TV picture.

Sparklies—Colored snow. Commonly used to refer to weak signal performance of satellite TV receivers.

Splicer—A device that joins two pieces of a single type of cable or conductor.

Splitter—See *signal splitter.*

Stereo (Stereophonic)—A sound reproduction system in which sound is delivered to the listener through at least two channels, left and right, creating the illusion of depth and locality of source.

Switch—A mechanical or electrical device that completes or breaks the path of the current or sends it over a different path.

Switcher Panel—A device that contains multiple signal switches and allows one of a number of signals to be selected for connection to a device or one of a number of devices to be connected to a signal. Also see *video switcher.*

Television (TV)—A telecommunication system for transmitting images of fixed or moving objects.

Terminator Plug—A special connector containing a resistor that supplies a proper impedance match for an unused port on a signal splitter or other device with multiple outputs. (Without the terminator, signals reaching the open port can be reflected and possibly interfere with the incoming signal.)

Transmission Line—Twin lead or coax cable from an antenna or cable TV system.

Trap—A type of interference filter that eliminates a single frequency or band.

Tuner—That portion of a TV set, VCR, cable box, or other accessory that selects a specific VHF, UHF, or cable channel.

TV Stereo Audio Decoder—An accessory that connects to a TV set to receive and decode TV stereo sound.

Twin Lead—A transmission line with two conductors held parallel by insulation. A flat cable with an impedance of 300 ohms used for VHF and UHF TV signals.

UHF (Ultrahigh Frequency)—TV channels 14-83, frequency band 300 MHz to 3 GHz, wavelength 1 m to 10 cm.

UHF Up Converter—A device for changing the entire UHF band to a selected range of UHF channels.

Unscrambled Channel—A standard RF TV signal that can be played without a decoder.

Up Converter—A device for changing the signals of a band of frequencies to the same signals at a higher band of frequencies.

VHF (Very High Frequency)—TV channels 2-13 and cable channels AA-WW; frequency band 30 to 300 MHz, wavelengths 10 m to 1 m.

VHS—A recording format for ½-inch video tape. The most popular VCR format.

Video—Picture signals. The signal originally produced by a TV camera.

Video Cassette Recorder (VCR)—A consumer electronic product that records and plays back audio and video signals on magnetic tape held in plastic cassettes.

Video Control Center—See *video switcher* and *switcher panel.*

Videodisc Player—A device that reproduces audio and video signals stored on a mass-produced platter.

Video Processor—An electronic accessory used to enhance or correct the color signals when recording or playing back video on a VCR.

Video Switcher—An accessory box to which all video accessories (TV set, VCR, and so on) are connected for the purpose of quickly and conveniently connecting or disconnecting them with front panel switches. Also see *switcher panel.*

Video System—All the equipment, accessories, wires, and signal sources that allow you to tune, record, and view TV programs.

Index

Compact Disc Troubleshooting & Repair

Neil R. Heller

This book discusses much more than how to repair a compact disc, it is a concise and complete introduction to the field of digital signal processing and compact disc players. The book describes in detail the hardware and technology involved in this expanding new field.

Topics covered include:

- The Audio Signal and Signal Processing
- Basic Digital Theory
- The CD Modulation System
- Construction of the Compact Disc
- The Laser Pickup
- Detailed CD Circuit Descriptions
- Adjustments and Troubleshooting
- The CD-ROM
- Glossary of Terms

160 Pages, 8½ x 11, Softbound
ISBN: 0-672-22521-2
No. 22521, $19.95

Commodore 64® Troubleshooting & Repair Guide

Robert C. Brenner

Repair your Commodore 64 yourself, simply and inexpensively. Troubleshooting flowcharts let you diagnose and remedy the probable cause of failure. A chapter on advanced troubleshooting shows the more adventuresome how to perform complex repairs. Some knowledge of electronics is required.

Topics covered include:

- Theory of Operation and an Overview of System Structure
- Preventive Maintenance Tips
- Basic and Advanced Fix-It-Yourself Procedures
- Repair with Few or No Tools
- Complete Appendices of All Necessary Technical Data

256 Pages, 8½ x 11, Softbound
ISBN: 0-672-22363-5
No. 22363, $19.95

Commodore® 1541 Troubleshooting & Repair Guide

Mike Peltier

This guide provides troubleshooting and repair procedures for the Commodore 64® or VIC 20™ with a 1540, 1541, or 1542 disk drive. It is comprehensive and detailed, offering easy-to-follow, step-by-step instructions. It presents the theory and general operation of the disk drive and points out some common disk drive problems. The diagrams will help readers isolate their disk drive problems.

Easy-to-understand circuit diagrams and simple instructions written in a clear, concise style allow you to make most repairs with few or no tools.

Topics covered include:

- Description of the VIC-1541
- Theory of Operation
- Alignment
- Preventive Maintenance
- Basic Troubleshooting and Repair
- Advanced Troubleshooting and Repair
- Advanced Theory of Operation
- Appendices: Technical Data, Fabrication of Video Detector, MOS Handling Precautions, Standard IC Pin Numbers, Description of Disk Controller IC

200 Pages, 8½ x 11, Softbound
ISBN: 0-672-22470-4
No. 22470, $19.95

How to Maintain and Service Your Small Computer

John Stephenson and Bob Cahill

Simple repairs can be diagnosed without a trip to the repair store. This book shows you some easy maintenance and operating procedures that will likely reduce problems and down-time. If the problem is more major, then you can apply the skill learned here when dealing with the repair shop for professional servicing. Some basic electronic knowledge is required.

Topics covered include:

- Today's Equipment—A Service Profile
- Servicing Your Equipment—the Options
- Environment and Operating Procedures—Reducing Potential Problems
- Preparation—Skills, Documentation, and Tools
- Troubleshooting Strategy and Skills
- Troubleshooting
- Disk Drives
- Servicing Printers
- Troubleshooting the Main Computer and Keyboard
- Modems, Monitors, Cassette Recorders, and Game Controllers
- Troubleshooting Power Supplies
- Appendices: Resistor Color Code, IC Manufacturers' Numbering Systems, Mail-Order Parts Dealers, Pinouts and Base Diagrams, and Troubleshooting Chart

224 Pages, 8½ x 11, Softbound
ISBN: 0-672-22016-4
No. 22016, $17.95

Visit your local book retailer, use the order form provided, or call 800-428-SAMS.

John D. Lenk's Troubleshooting & Repair of Microprocessor-Based Equipment

John D. Lenk

This text provides a simplified, practical system of troubleshooting for the many types of microprocessor-based electronic devices. Concentrating on the basic approach to troubleshooting, the book includes numerous procedures and "tricks" that can be effective in diagnosing, isolating, and locating faults in microprocessor circuits.

Topics covered include:

- Microprocessor Test Equipment
- Problems in Troubleshooting Microprocessors
- Troubleshooting the Gate and IC
- Stimulus-Response Testing
- Vcc and Ground Shorts from Unexpected Causes
- Flip-Flop Troubleshooting
- Register Troubleshooting
- Current Flow Analysis
- Wired-AND and Wired-OR Troubleshooting
- Using the Logic Analyzer
- Step-by-step Procedure for Troubleshooting the VCR

John D. Lenk is one of the industry's foremost authorities on electronics and troubleshooting.

176 Pages, 8½ x 11, Softbound
ISBN: 0-672-22476-3
No. 22476, $21.95

Printer Troubleshooting & Repair

John Heilborn

The computer hobbyist with a basic understanding of electronics can learn to service, troubleshoot, and repair a printer with this easy-to-use guide. It includes all types of printers—dot matrix, ink jet, and laser.

Preventive maintenance is emphasized with common malfunctions and repair procedures highlighted. Well-illustrated coverage is complete with troubleshooting diagrams and detailed drawings.

Topics covered include:

- A Look at Printer Types: How They Work
- Troubleshooting Techniques
- Print Heads
- Ribbon, Paper, and Print Head Transport Mechanisms
- Power Systems
- Logic and Control Systems
- Laser Printers
- Preventive Maintenance
- Appendices: IC Pin Diagrams and List of Suppliers

John Heilborn is a computer book author in the San Francisco Bay Area with approximately 20 books to his credit.

200 Pages, 8½ x 11, Softbound
ISBN: 0-672-22543-3
No. 22543, $19.95

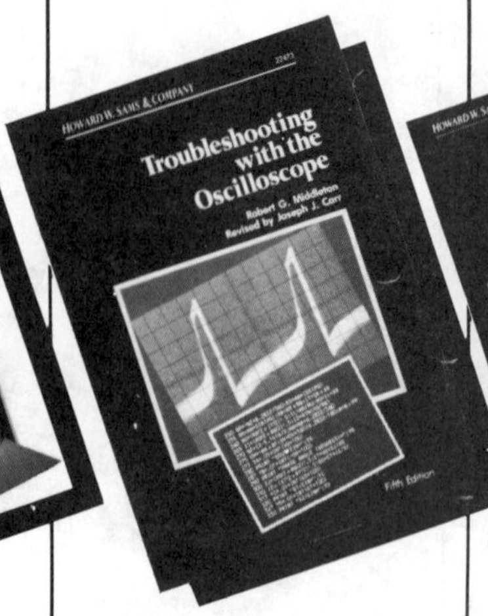

Troubleshooting with the Oscilloscope, Fifth Edition

Robert G. Middleton,
Revised by Joseph J. Carr

This 225-page book will help technicians and hobbyists obtain the maximum benefits from the wide selection of scopes now available for servicing and repair.

The new edition emphasizes digital troubleshooting and applications as well as the time- and data-domain analyzers and the logic state analyzer, with up-to-date information about television receiver troubleshooting.

Topics covered include:

- Introduction to the Oscilloscope
- How to Operate an Oscilloscope
- Using Oscilloscope Probes
- TV Signal Tracing in RF, IF, and Video Amplifiers
- Signal Tracing in the TV Sync Section
- Troubleshooting in the TV AFC and Horizontal Oscillator Section
- Waveform Tests in the Horizontal Sweep Section
- Troubleshooting in the TV Vertical Sweep Section
- Signal Tracing in the TV Sound-IF and Audio Sections
- Special Oscilloscopes and Their Applications
- Digital Logic Troubleshooting with the Oscilloscope

225 Pages, 5½ x 8½, Softbound
ISBN: 0-672-22473-9
No. 22473, $16.95

John D. Lenk's Troubleshooting & Repair of Audio Equipment

John D. Lenk

This manual provides the most up-to-date data available and a simplified approach to practical troubleshooting and repair of major audio devices. It will enable both the beginning and the intermediate level technician or hobbyist to apply tips and tricks to any specific equipment.

This book also includes such time-saving hints as circuit-by-circuit troubleshooting based on failure or trouble symptoms, universal step-by-step procedures, and actual procedures recommended by manufacturers' service personnel.

Topics covered include:

- Introduction to Modern Audio Equipment Troubleshooting
- Troubleshooting and Repair of Amplifiers and Loudspeakers
- Troubleshooting and Repair of Linear-Tracking Turntables
- Troubleshooting and Repair of Audio Cassette Decks
- Troubleshooting and Repair of AM/FM Stereo Tuners
- Troubleshooting and Repair of CD Players

John D. Lenk is the author of over 60 books in electronics. Over one million of his books have been sold in six different languages.

208 Pages, 8½ x 11, Softbound
ISBN: 0-672-22517-4
No. 22517, $21.95

Visit your local book retailer, use the order form provided, or call 800-428-SAMS.

IBM® PC Peripheral Troubleshooting and Repair

C. J. Brooks

Technically oriented personal computer owners, service technicians, students, and hobbyists can learn to service and repair computer peripherals with this step-by-step procedures manual.

Using a cookbook approach, this book presents the basic theory of operation, hardware descriptions, preventive maintenance, and troubleshooting and repair procedures for IBM computer peripherals. Flowcharts, block diagrams, lists of replacement parts, and many illustrations make basic repairs clear.

Reference is made to COMPUTERFACTS™ technical service data from Howard W. Sams when relevant.

Topics covered include:

- Basic Computer Organization
- Troubleshooting Basics
- Disk Drives/Hard Disk Drives/Floppy Disk Drives
- Monitors: IBM Monochrome
- Input Devices: Keyboards, Keypads, and Pointing Devices
- Printers/Impact Printing/Non-Impact Printing
- Modems

320 Pages, 8½ x 11, Softbound
ISBN: 0-672-22529-8
No. 22529, $19.95

Apple® II Plus/IIe Troubleshooting & Repair Guide

Robert C. Brenner

This manual is structured to make it quick and easy for either the novice or the experienced technician to locate and correct most computer failures.

It introduces troubleshooting and repair in easy-to-understand terms and proceeds step-by-step through the diagnostic techniques for microcomputer hardware.

The appendices provide a wealth of back-up information including step-by-step disassembly and reassembly instructions, conversion tables, specification data sheets, and system configuration record and repair sheets.

Topics covered include:

- Basic Troubleshooting
- The Apple II Plus Described
- Apple II Plus Operations
- Specific Troubleshooting and Repair, Apple II Plus
- Apple IIe Described
- Apple IIe Operations
- Specific Troubleshooting and Repair, Apple IIe
- Routine Preventive Maintenance
- Advanced Troubleshooting

256 Pages, 8½ x 11, Softbound
ISBN: 0-672-22353-8
No. 22353, $19.95

IBM® PC Troubleshooting & Repair Guide

Robert C. Brenner

Keep your IBM PC in top operating condition with this handy reference. Its pages of schematics, photos, and block diagrams will help to identify problems and zero-in on the malfunctioning component quickly, for reduced down-time.

Easy-to-understand circuit diagrams and simple instructions written in a clear, concise style allow you to make most repairs with few or no tools.

Technicians, hobbyists and even computer novices will feel comfortable with the complex world of electronic troubleshooting. Discover how learning can be interesting and rewarding.

Topics covered include:

- The IBM PC Described
- IBM PC Operations
- Basic Troubleshooting
- Specific Troubleshooting & Repair for the IBM PC
- Preventive Maintenance
- Advanced Troubleshooting Techniques
- Appendices: IBM PC Data Sheet, Chip Information Chart, System Unit Disassembly Instructions, System Unit Reassembly Instructions, Keyboard and Disk Drive Disassembly and Reassembly Instructions, Replacing Surface-Mounted Components on the System Board, Summary of Cautions and Notes, ASCII Code Chart, and a Hexadecimal-to-Decimal Conversion Chart

208 Pages, 8½ x 11, Softbound
ISBN: 0-672-22358-9
No. 22358, $19.95

How to Read Schematics, Fourth Edition

Donald E. Herrington

This update of a standard reference features expanded coverage of logic diagrams and a chapter on flow charts. Beginning with a general discussion of electronic diagrams, the book systematically covers the various components that comprise a circuit. It explains logic symbols and their use in digital circuits, interprets sample schematics, analyzes the operation of a radio receiver, and explains the various kinds of logic gates.

Topics covered include:

- Types of Electronic Diagrams
- Resistors
- Capacitors
- Coils and Transformers
- Electron Tubes
- Semiconductor Diodes and Related Devices
- Transistors and Integrated Circuits
- Switches and Relays
- Miscellaneous Components
- Connecting the Components
- Combining the Components
- Flowcharts
- Digital Circuits
- Reading and Interpreting Schematic Diagrams
- Appendices: House Wiring Symbols, Appliance Symbols, Answers to Questions

272 Pages, 5½ x 8½, Softbound
ISBN: 0-672-22457-7
No. 22457, $14.95

Visit your local book retailer, use the order form provided, or call 800-428-SAMS.

Commodore 128® Troubleshooting & Repair

John Heilborn

This book is a step-by-step "how-to" manual written for the owner or operator of a Commodore 128. From hobbyist to service technician, users at all levels will find the information necessary for properly maintaining and repairing the C-128 computer.

Another in the successful consumer-oriented troubleshooting and repair series for Sams, this book covers diagnostic techniques, operation, hardware, preventive maintenance, required tools, data sheets, assembly and disassembly instructions, and a comprehensive glossary. The book explains the inner workings of the Commodore 128 and shows you how to avoid problems through proper care and maintenance.

Topics covered include:

- A Look at the Commodore 128
- Commodore 128 Systems and Subsystems
- Troubleshooting Techniques
- Digital Logic Circuits
- Troubleshooting a System That Doesn't Work
- Troubleshooting a Poor Video or No Video
- Troubleshooting a Defective Keyboard
- Troubleshooting Input/Output Problems
- Preventive Maintenance
- Appendices: Integrated Circuit/Pinouts, Board Layout, I/O Ports, System Specifications, and Commodore Keyboard Codes

160 Pages, 8½ x 11, Softbound
ISBN: 0-672-22554-9
No. 22554, $19.95

VCR Troubleshooting & Repair Guide

Robert C. Brenner

With approximately 25 million video cassette recorders on the market, this long-awaited book will help owners repair these popular machines when they break down.

This helpful troubleshooting guide is for the electronics hobbyist, layperson, or technician who needs a preventive maintenance and troubleshooting reference for VCRs. Limited electronics experience is required to use it, but more sophisticated service and repair functions are discussed, and valuable information for the service technician is included.

Topics covered include:

- Introduction
- Principles of Video Recording
- The VCR System Described
- Basic VCR Troubleshooting Techniques
- Troubleshooting Specific Problems
- VCR Preventive Maintenance
- Appendix and Glossary of VCR Terms

Robert Brenner is a specialist in microelectronic applications as well as troubleshooting and repair of electronic equipment. He is also the author of the best-selling troubleshooting and repair books for the Apple,® IBM,® and Commodore 64.®

256 Pages, 8½ x 11, Softbound
ISBN: 0-672-22507-7
No. 22507, $19.95

How to Read and Interpret Schematic Diagrams

J. Richard Johnson

This book teaches step-by-step the recognition of schematic symbols, their use and function in diagrams, and the interpretation of these diagrams for design, maintenance, and repair of electronic equipment.

The author takes the reader logically through the fundamentals of schematic diagrams including component symbols and diagram formation and functional flow and block diagrams.

Topics covered include:

- The What and Why of the Schematic Diagram
- Component Symbols and Diagram Formation
- Functional Sequence and Block Diagrams
- Power Supplies: Active Devices and the Need for a Power Supply
- Audio System Diagrams
- Radio Receivers
- Radio Transmitter and Transceiver Diagrams
- Interpreting Television Receiver Diagrams
- Computer Diagrams

210 Pages, 6 x 9, Softbound
ISBN: 0-8104-0868-6
No. 40868, $12.95

Crash Course in Electronics Technology

Louis E. Frenzel, Jr.

This third book in the Howard W. Sams "Crash Course" series will help hobbyists, technicians, students, and laypersons learn the basics of electricity and electronics.

In a step-by-step format, electricity is introduced briefly followed by a discussion of circuit basics and electronic devices. The book continues with electronic communication, controls, motors, test equipment, and troubleshooting. Everything is presented in a self-paced, self-instructional format.

Topics covered include:

- Introduction to Electronics
- Principles of Magnetism and Electricity
- Resistors and Resistive Circuits
- Inductance and Transformers
- Capacitance and Capacitors
- Resonance and Filters
- Diodes and Transistors
- Amplifiers
- Operational Amplifiers
- Oscillators
- Pulse Techniques
- Power Supplies
- Amplitude and Frequency
- Industrial Control
- Test Equipment and Troubleshooting

400 Pages, 8½ x 11, Softbound
ISBN: 0-672-22494-1
No. 22494, $21.95

Visit your local book retailer, use the order form provided, or call 800-428-SAMS.